Contents

Physics continued

Introduction

Take advantage of the CD

Cambridge Essentials Science comes with a CD in the back. This contains the entire book as an interactive PDF file, which you can read on your computer using free Acrobat Reader software from Adobe (http://www.adobe.com/products/acrobat/readstep2.html). As well as the material you can see in the book, the PDF file gives you extras when you click on the buttons you will see on most pages; see the inside front cover for a brief explanation of these.

To use the CD, simply insert it into the CD or DVD drive of your computer. If you are using Microsoft Windows, you will be prompted to install the contents of the CD to your hard drive. Installing will make it easier to use the PDF file, because the installer creates an icon on your desktop that launches the PDF directly. However, it will run just as well directly from the CD.

If you are using an Apple Mac computer or any other operating system, or you simply want to copy the disc onto your hard disc yourself, this is easily done. Just open the disc contents in your file manager (for Apple Macs, double click on the CD icon on your desktop), select all the files and folders and copy them wherever you want.

Take advantage of the web

Cambridge Essentials Science lets you go directly from your book to web-based activities on our website, including animations, exercises, investigations and quizzes. Access to these materials is free to all users of the book.

There are three kinds of activity, each linked to from a different place within each unit.

- **Scientific enquiry:** these buttons appear at the start and end of each unit. The activities in this section allow you to develop skills related to scientific enquiry, including experiments that would be hard to carry out in the classroom.
- **Check your progress:** these buttons come half-way through each unit. They let you check how well you have understood the unit so far.
- **Review your work:** these buttons come at the end of each unit. They let you show that you have understood the unit, or let you find areas where you need more work.

The *Teacher Materials* CD-ROM for *Cambridge Essentials Science* contains enhanced interactive PDFs. As well as all the features of the Pupil PDF, teachers also have links to the *Essentials Science* Planner – a new website with a full lesson planning tool, including example lesson plans, worksheets, practicals, assessment materials and guidance. The e-learning materials are also fully integrated into the Planner, letting you see the animations in context and alongside all the other materials.

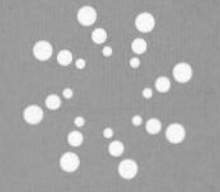

9A.1 Variation and its causes

Scientific enquiry

You should already know

Outcomes

Keywords

Living organisms vary

Different species of plants and animals look different. Even within the same species, animals and plants have differences. We say that they show **variation**.

Non-identical twins.

Different varieties of tomatoes.

Question 1 2

The things you described in your answers to questions 1 and 2 are called **characteristics**. Many characteristics are passed from parents to offspring. We say they are **inherited**.

Identical twins inherit the same characteristics from their parents.

Question 3

Is all variation inherited?

It's hard to tell the difference between the young identical twins in the picture.

But the adult identical twins are not exactly the same. They inherited the same characteristics, but different things have happened to them.

The environment has caused differences. This is called **environmental variation.**

These adults are identical twins but they show environmental variation.

Question 4

Identifying causes of variation

Many characteristics pass from generation to generation. Look at the picture.

Four generations: great-grandmother, grandmother, mother and baby.

Question 5

Tomato colour is passed from generation to generation. It is inherited. So yellow tomatoes come from plants grown from seeds of yellow tomatoes. But the reasons for variation in size are more complicated.

Question 6 7

A August 4th **B** August 12th

On August 4th, tomato A is bigger than tomato B.
On August 12th, they are both fully grown and are the same size.

Sometimes tomato plants with identical inherited characteristics are different. The environment causes these differences.

> **Remember**
> To grow, plants need sunlight, water, carbon dioxide and minerals.

Question 8

Patrick's experiment

I grew some plants using seeds from one plant. I had some problems. There were so many plants that I had to keep some next to the window and some on a shelf. Sometimes I forgot to water them.

After 6 weeks, I measured the heights of the plants.

Height of plant (cm)	Number of plants
31–40	6
41–50	10
51–60	15
61–70	12
71–80	8

Patrick's results.

Question 9 10 11

9A.2 Why are individuals similar but not identical?

You should already know | Outcomes | Keywords

A complicated set of chemical instructions is needed to build your body. Your body cell nuclei contain these instructions. Each nucleus contains 23 pairs of **chromosomes** (that is, a total of 46). Small sections of these chromosomes are called **genes**. Each gene carries the instructions for a particular characteristic. We say that genes contain **genetic information**.

So genes are coded instructions.

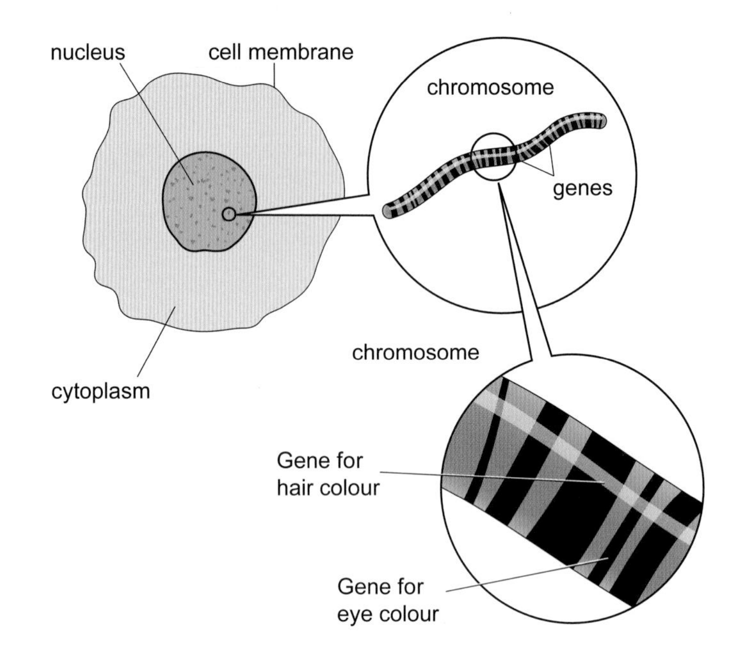

Chromosomes in your cell nuclei contain genes that code for characteristics such as hair colour and eye colour.

Question 1 **2**

You have learned that the human sex cells are sperm from a man and egg cells from a woman. Sperm and egg cell nuclei each contain 23 chromosomes. The diagram shows how these nuclei are made.

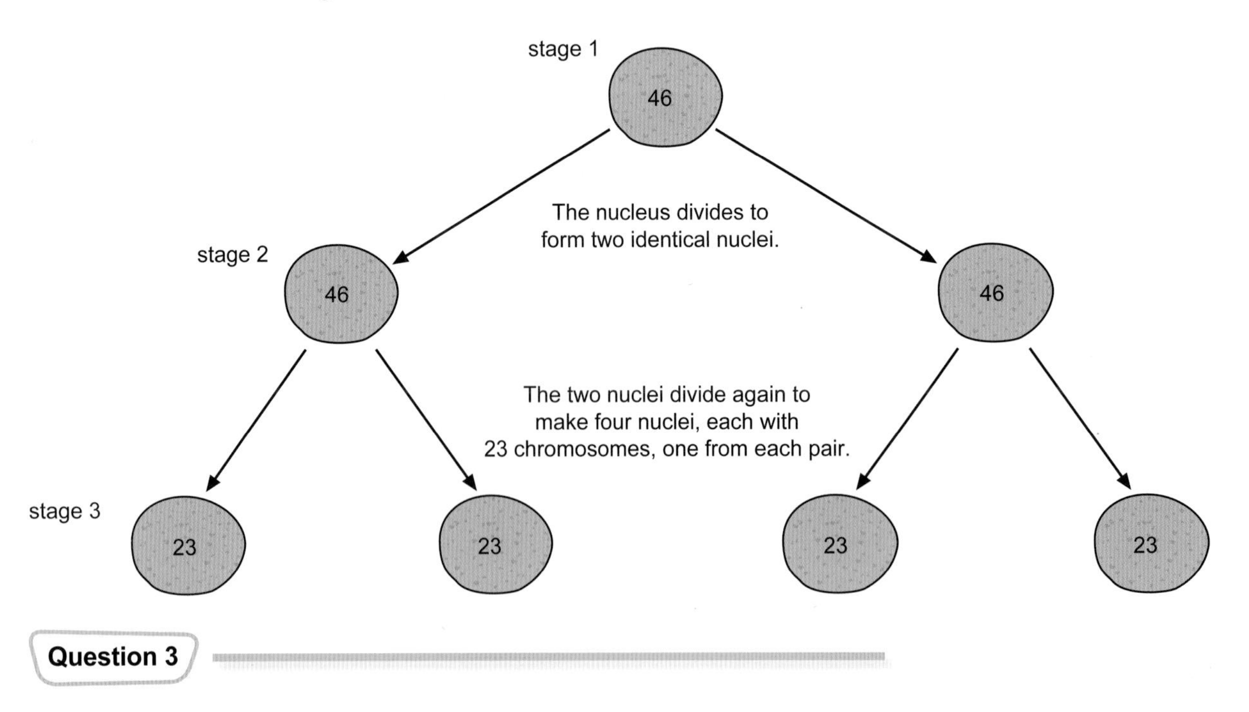

Question 3

Passing on the instructions

For a new life to begin, a nucleus from a sperm must join with a nucleus from an egg cell. This is called **fertilisation**.

The diagram shows the nucleus of a sperm joining with the nucleus of an egg cell. The new cell contains genetic information from both sex cells.

This new cell is unique because it contains some information from the mother and some from the father.

The new life will have a selection of characteristics from both parents. So sexual reproduction results in variation.

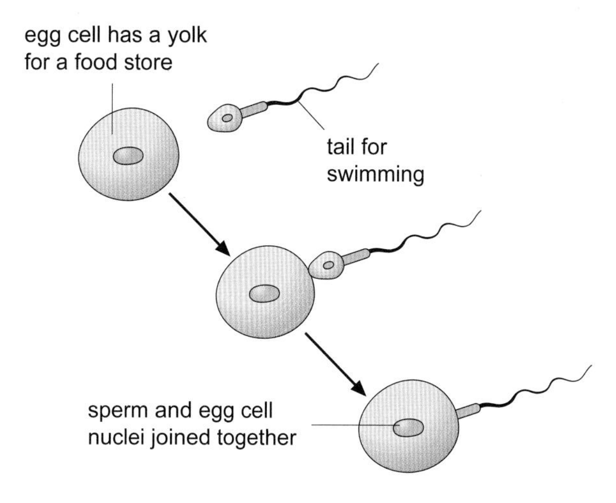

Fertilisation in animals.

Question 4 

Fertilisation in plants

Sexual reproduction in plants also involves a male and a female nucleus joining together. These nuclei also contain genetic information.

In flowering plants, male sex cells are inside pollen grains and female sex cells are inside ovules.

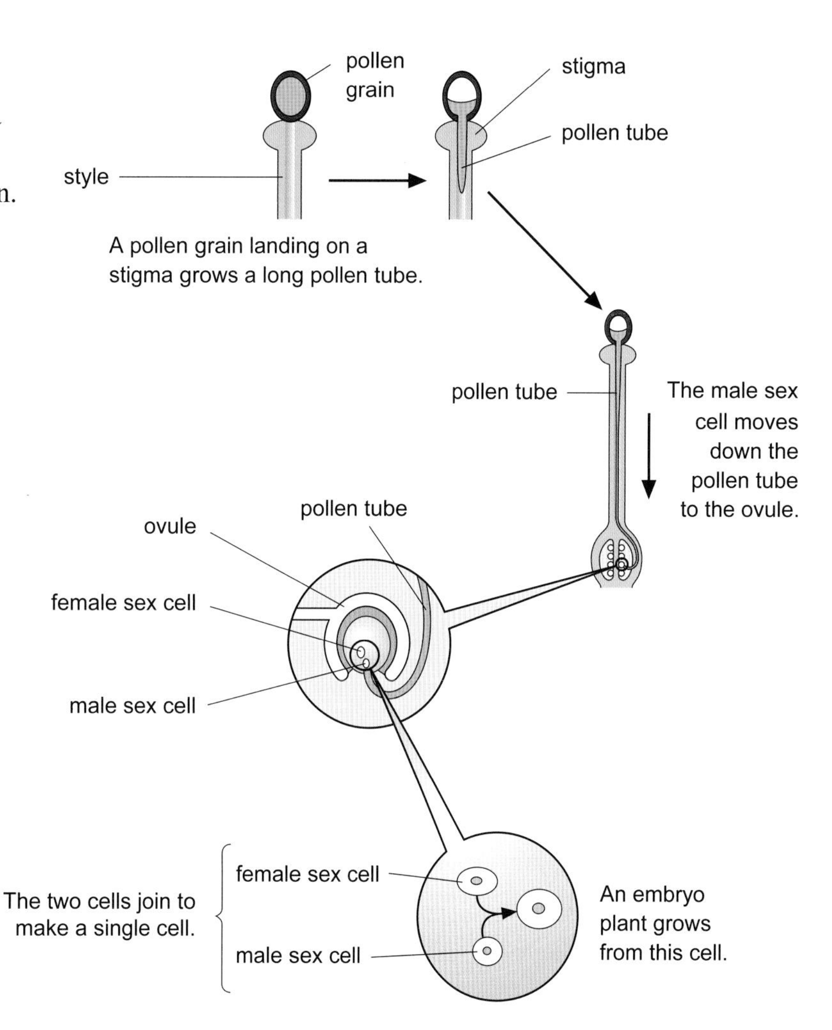

Fertilisation in flowering plants.

Question 8 

9A.3 The right breed for the job

You should already know | Outcomes | Keywords

Even though they look different, the animals in the pictures belong to the same species. Breeders have selected dogs with certain characteristics and allowed them to breed. Some examples of characteristics they have selected are size, colour and thickness of coat.

Sheepdog.

Dachshund.

Question 1  2

Selecting individuals with the most suitable characteristics and breeding from them is called **selective breeding**. There is evidence that humans domesticated a few wolves in Asia tens of thousands of years ago. Over the millennia, they developed all the different varieties of dogs from just these few wolves by selective breeding.

How selective breeding works

In nature, animals choose their own mates. The genes for lots of different characteristics are mixed over and over again. It's a bit like shuffling cards and dealing them out. You never know which characteristics will appear in each hand of cards.

In selective breeding, humans decide which animals will mate. Humans choose animals with the characteristics that they want. Think of it like choosing your cards instead of relying on chance.

Choose, from the animals that they have, male and female animals with the characteristics that they want.

↓

Breed from these animals.

↓

Select the offspring with the required characteristics and breed from them.

↓

Repeat for several generations.

What breeders do.

Question 3

Farmer Mansfield's goats

Farmer Mansfield wants to breed goats with longer, softer coats.

Here are some of the farmer's female goats that can be used for the breeding.

Some of Farmer Mansfield's female goats.

Question 4 5 6 7 8 

Why humans produce new breeds

Selective breeding benefits humans in many ways. For example, Farmer Mansfield expects to be able to get more money for the hair from his goats if it is softer and longer.

Breeders select cows that produce more milk.

Breeders select dogs for a variety of uses, such as guard dogs, guide dogs, sheepdogs and cute pets.

Breeders select beef cattle with big muscles to produce lots of tasty, tender beef.

Breeders select hens that lay more eggs.

Question 9 10

Check your progress

9A.4 From selective breeding to selection in nature

You should already know | Outcomes | Keywords

Body structure and chemistry are not the only characteristics affected by breeding. Behaviour is affected too. This is because some behaviour, including some social behaviour, is **innate** (**instinctive**). It is determined by the animal's **genes**.

In nature, animals are more likely to survive and breed if their behaviour is suited to their environment and way of life.

Each species has its own strategy. You have learned that some animals group together only when they are ready to mate. Many animals live in groups all the time. These social groups range in size from small family groups to colonies of many thousands.

The table compares some examples of social and **breeding behaviour** in humans and other animals. For survival and for successful breeding, an animal's behaviour has to fit in with the patterns for its species.

Remember

- Genes contain genetic information.
- Individuals within a species vary.
- We can breed from living things with the genes (characteristics) that we want (selective breeding).
- Sex hormones influence breeding behaviour.

Behaviour …	… in humans	… in other animals
Forming a pair bond.	Courting and marriage.	Many male birds 'show off' their plumage, singing or nest site, or their food-gathering or nest-building abilities to attract a mate.
Care of young.	Parental and other care and teaching for a long time.	Anything from no care at all to long-term care and teaching.
Territorial behaviour and aggression.	Groups, households and countries have their rules, laws and morals. May use aggression to gain territories or resources.	Breeding territories, permanent territories, no territories. Behaviour patterns often limit damage or death.
Pecking order.	Leadership structures in a group.	Weaker animals give way to dominant ones.
Sharing work.	Various patterns according to culture in both workplace and home.	Various patterns according to species – from none to total co-operation, e.g. a wolf pack hunting.

Some examples of behaviour in humans and other animals.

Question 1 2 3

Birds in a flock squawk, fluff out their feathers and peck at each other to sort out who is 'top bird', second bird and so on.
This is where the term 'pecking order' comes from.

Animals competing for mates sometimes wound each other. But usually one gives up before serious harm is done.

Question 4 5

Survival of colonies

The bee in the middle has filled the pollen sacs on its legs. It does a 'waggle dance' to show other workers where to find food.

Colonies of animals such as ants and bees depend on sharing work and co-operative behaviour for survival.

The job of a queen bee in a hive is to lay eggs. The drones are the males that mate with the queen so that eggs can be fertilised.

Worker bees are sterile females. They live for about 30 days after emerging from their cells. Most of their behaviour is innate. Their genes programme their physical development and their changes in behaviour. Look at the table.

Day 1	Stage of development	Job or behaviour
	newly emerged worker	cleaning brood cells and honeycomb
	glands in the throat start to make a special food called royal jelly	feeding and grooming larvae
	wax-secreting glands active	building new cells
	learned the smells of the hive	receiving and storing nectar and pollen
	gained strength and learned to recognise intruders	guarding and regulating temperature of the hive by fanning their wings
Day 30	learned the communication system (waggle dance) and how to find their way back to the hive	foraging (finding and gathering nectar and pollen)

Remember
Scientists found out about the behaviour of bees by observing them. Karl von Frisch is the scientist famous for working out how the 'waggle dance' showed both the direction and distance of food from the hive.

The science of observing the behaviour of animals in their natural surroundings is called <u>ethology</u>.

Question 6  7

9A.5 Selective breeding of plants

You should already know | **Outcomes** | **Keywords**

Plant breeders are producing new varieties of plants all the time. The different varieties of tomatoes shown in Topic 9A.1 and the beans on this page were all produced by **selective breeding**.

Farmer Kirkby's bean plants varied.

This plant is tall and has a strong stem but produces only a small amount of fruit. It can survive in strong winds and cold weather.

This plant is small but produces very big fruits. Strong winds and cold weather damage it.

Question 1

Farmer Kirkby wanted tall, strong plants that yield lots of fruit. She produced them by selective breeding. She chose the plants that were most like the ones that she wanted and transferred pollen from one plant to the other. After several generations, she had bred offspring with the characteristics that she wanted.

Look at the diagram.

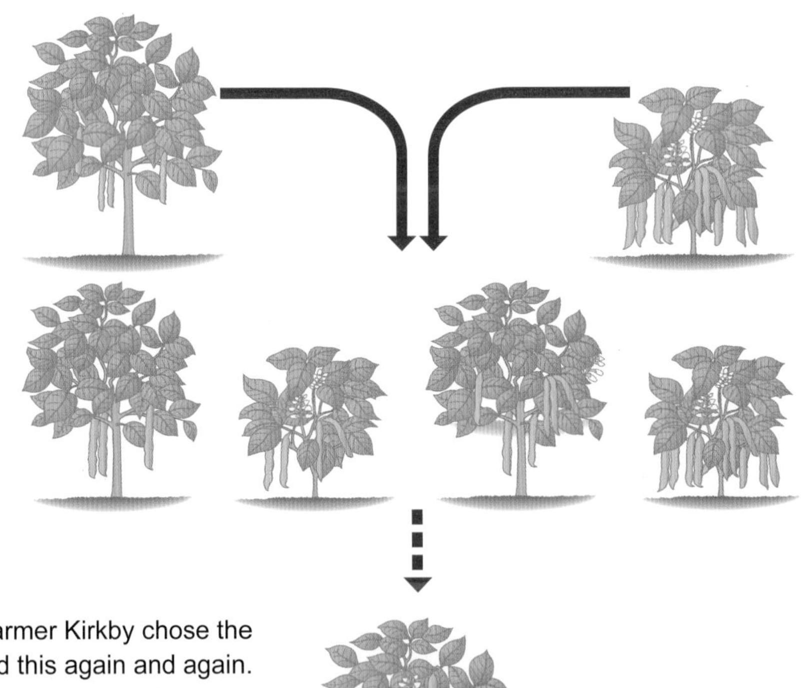

From these offspring, Farmer Kirkby chose the best to breed from. She did this again and again. Eventually, she produced plants like this one.

Question 2 **3**

Plant breeding

You have seen how pollen on a stigma grows a long pollen tube. Then the male sex cell goes down the tube to the ovule and joins with the female sex cell.

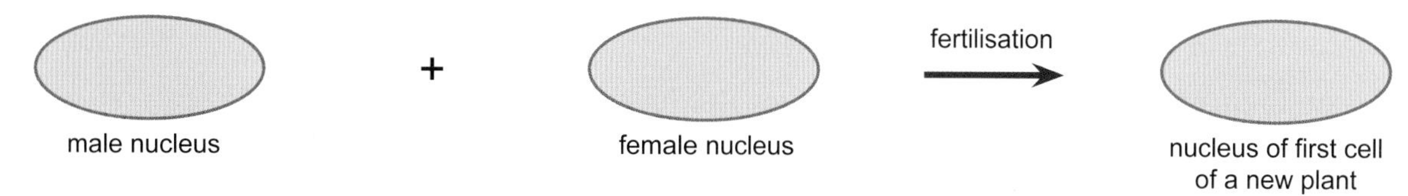

In sexual reproduction, the two nuclei join together. We call this **fertilisation**.

Plant breeders have to make sure that flowers are pollinated only with pollen from the plants that they have selected. It is important that plants are not pollinated by accident.

The diagram shows what plant breeders do.

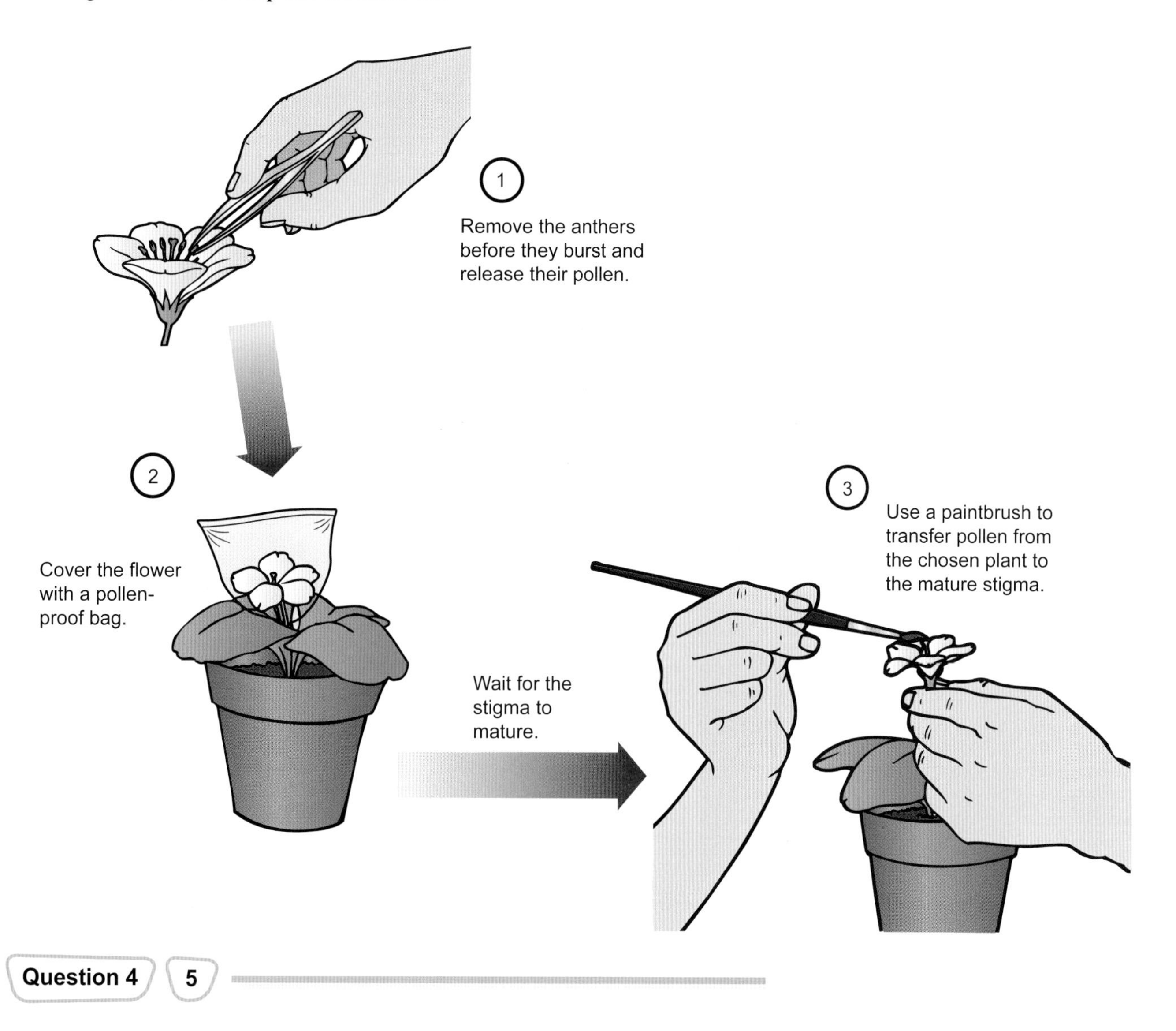

Question 4 5

9A.6 What is a clone?

You should already know

Outcomes

Keywords

Asexual reproduction is a type of reproduction in which only one parent is involved. The offspring have exactly the same genes as the parent.

We say they are genetically identical. They are called **clones**.

> **Remember**
> In **sexual reproduction** the nuclei of the sex cells from two parents join.

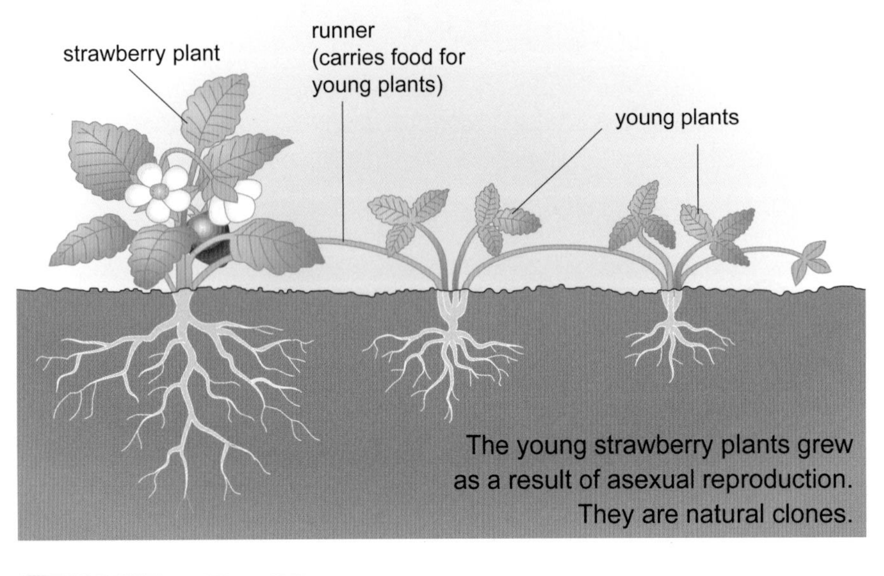

Question 1 2 3

Taking cuttings

Taking cuttings is a quick and cheap way of producing lots of new plants.

These new plants have exactly the same genes as the parent plant and as each other. So they are clones.

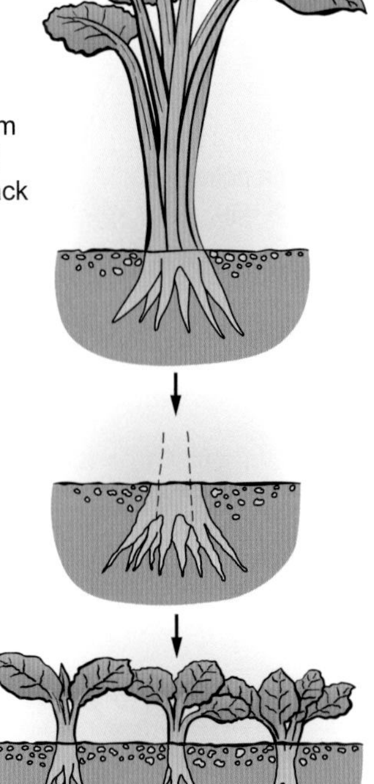

Question 4    5 6 7

Producing plants by grafting

Grafting is a method that gardeners use to reproduce trees. They join the cut stems of two plants with tape so that the tissues grow together.

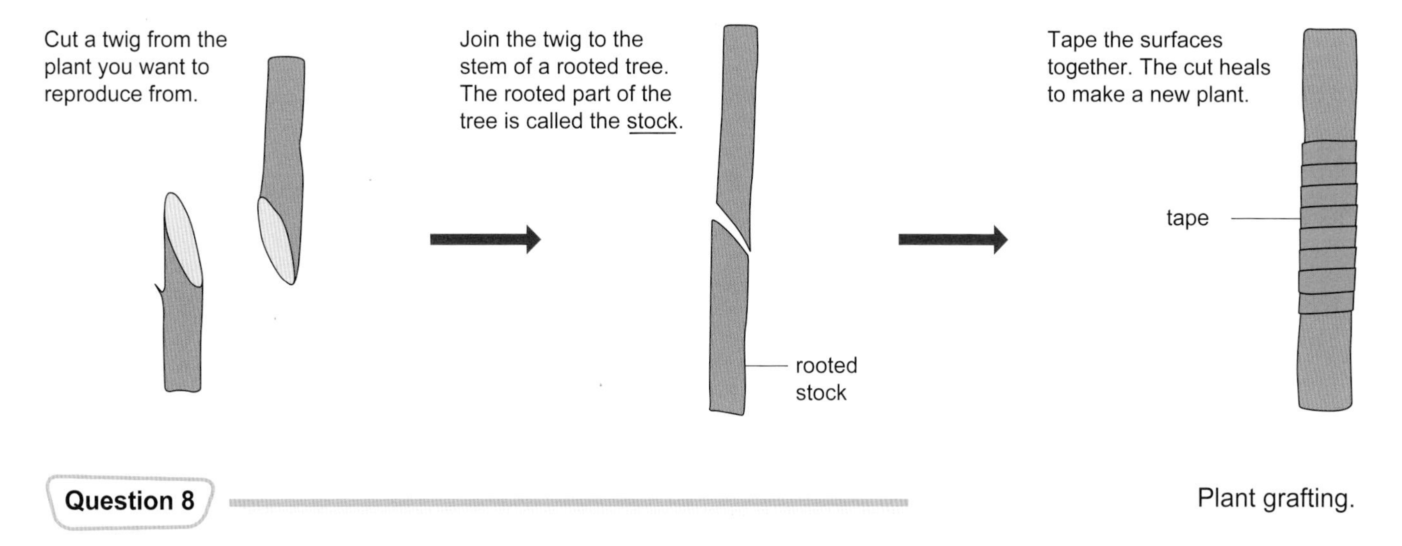

Plant grafting.

Question 8

Fruit trees produce seeds as a result of sexual reproduction. Trees grown from seeds vary.

If you plant an apple pip, it will be several years before the tree is big enough to produce apples. You will have no idea what the apples will be like.

Grafted fruit trees are clones, so their characteristics are exactly the same as those of the tree from which the graft was taken.

Question 9

Animal clones

Some tiny animals also reproduce asexually and produce clones. But sexual reproduction is normal in larger animals.

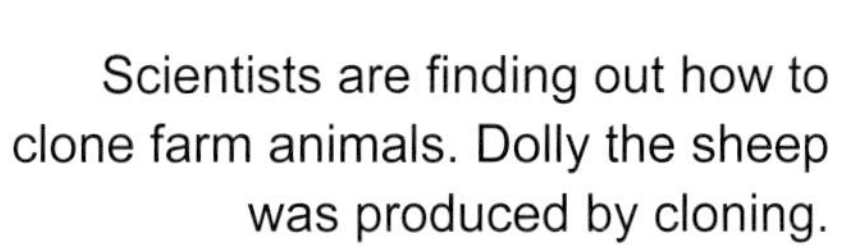

Scientists are finding out how to clone farm animals. Dolly the sheep was produced by cloning.

Question 10

Review your work

Summary ➡

9A.HSW Altering the genes – ethical issues

You should already know

Outcomes

Keywords

In this unit, you have seen how we can alter populations by **selective breeding**. We breed from plants and animals with the characteristics that we want. The questions on these two pages are for discussion in your group.

> **Remember**
> When you think about **ethics**, you need:
> - to understand what can be done;
> - to think of the advantages and disadvantages;
> - to decide what you think and give reasons.

Question 1

Selective breeding has its problems

When we increase the number of genes that are useful to us:

- other genes are lost, so there is less **variation** in the population;
- we may create problems for the plants and animals that we breed.

So, we need to ask ourselves about the ethics of doing these things.

Variation is important because some of the lost genes may be needed in the future. For example, suppose a lost gene helped the potato plant survive with less water. As the climate changes, we may need varieties of potato that are better able to survive water shortages.

This is one reason why some organisations are trying to preserve **biodiversity** – of both species and variety within a species.

In the UK, there is the Millennium Seed Bank. The Svalbard International Seedbank is in a secure tunnel in a mountain on the frozen island of Spitzbergen in Norway.

A laboratory at the Millennium Seed Bank building at Wakehurst Place, Sussex. The vault under this building already contains over a billion seeds.

Question 2   

We breed different dogs for all sorts of different reasons. Some breeds have severe health problems as a result of:

- the breed standard (the official list of what characteristics a breed should have);
- selective breeding that has resulted in some breeds having faulty genes.

Breed of dog	Suffers from …
Dachshund	back problems (breed standard)
German shepherd	hip problems (genetic)
Dalmatian	deafness (genetic)
Pekinese	breathing problems (breed standard)
Boxer	heart disease (genetic)

Question 6

Another way of altering plants and animals

Scientists have developed ways of taking genes out of one cell and putting them into another. This is called **genetic engineering**.

Some people agree and others totally disagree with doing this. Some think that it should be done only in certain circumstances.

Making hormones.

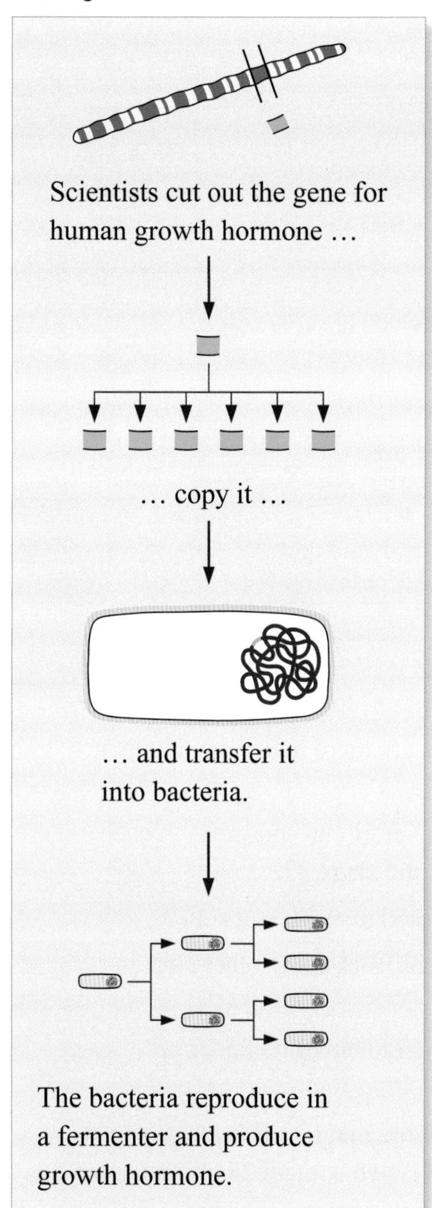

Curing diseases.

'Bubble babies' have no immune system because they can't make an enzyme called ADA.

Scientists …

… add the gene for the enzyme to harmless viruses …

… remove some of the baby's bone marrow cells…

… infect the cells with the viruses …

… put the cells back into the baby.

Sometimes the cells start to make the enzyme and the immune system works.

Protecting plants.

Some crops are attacked by insects.

Scientists …

… take a gene for making an insecticide (a substance that poisons insects) from a bacterium …

… and put it into the cells of crop plants such as corn …

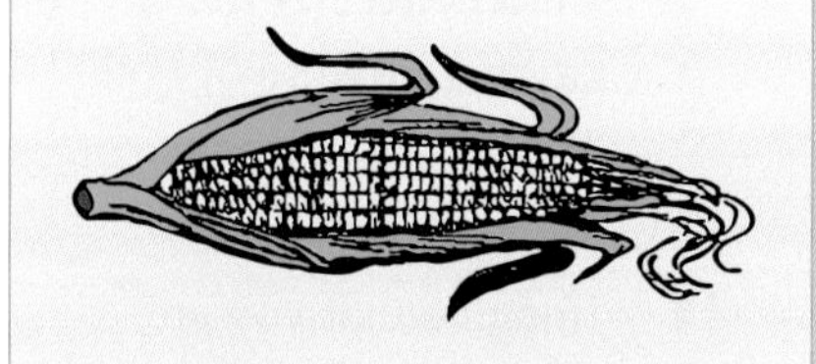

… where the plant cells make the poison. This kills the insects that eat the plant.

Question 7 8

9A Questions

9A.1

1 Write down two differences between the non-identical twins.

2 Sketch two tomatoes from the picture.
Label them to show two differences between them.

3 In tomatoes, suggest one characteristic not shown in the picture that may be passed from one generation to the next.

4 Write down two differences between the adult twins that are examples of environmental variation.

5 Look at the photograph of four generations of one family.
Write down:

a two inherited characteristics;

b two characteristics that you can see in the photograph that are affected by the environment.

6 What type of seed produces red tomatoes?

7 Suggest why tomato A is bigger than tomato B on August 4th, but they are the same size on August 12th.

8 Write down two environmental conditions that can affect how well a plant grows.

9 Draw a bar chart of Patrick's results.

10 The differences between Patrick's plants could have been caused by inheritance, by the environment or by a mixture of both.
What do you think?
Explain your answer.

11 Imagine that you are Patrick's teacher.
Write him a note explaining how to make his experiment a fair test.

9A.2

1 Which part of the cell contains the genetic information?

2 How many chromosomes are there in a normal human body cell nucleus?

3 What happens to the number of chromosomes in each nucleus between stage 2 and stage 3?

4 How are sperm cells adapted to reach egg cells?

5 How are egg cells different from many other cells in the body?

6 Why is a child not identical to either its father or its mother?

7 Use the information from the diagram to explain why a sperm cell and an egg cell each contain 23 chromosomes, but the cell after fertilisation contains 23 pairs of chromosomes.

8 How is fertilisation in plants similar to fertilisation in animals?

9 Explain why the plant that grows from each embryo is unique.

9A.3

1. Look at the pictures.
   Write down two ways in which the dogs are:
   - **a** alike;
   - **b** different.
2. **a** Which of these dogs might hunters and poachers use to bring animals out of their burrows?
   b Write down two characteristics of the dog that would make it suitable for this job.
3. When animals breed, a sperm cell from the male fertilises an egg cell in the female.
   What part of a cell contains the genes that control an animal's characteristics?
4. Look at the picture of the female goats.
   Which two goats will Farmer Mansfield let the male mate with?
5. How will the farmer decide which of the baby goats to keep for breeding and which to sell?
6. What is the next step in the breeding programme? Explain your answer.
7. Why do breeders make sure that both the male and female animals have the required characteristics before they allow them to breed?
8. Explain why breeders try to control the environmental conditions for the breeding animals.
9. Look at the pictures.
   Choose two examples and explain why the characteristics of each are useful for consumers.
10. What characteristics do you think that farmers try to develop in:
    - **a** apples?
    - **b** sheep?

9A.4

1. Suggest two ways that pair bonding improves the chances of survival of a pair's young.
2. Humans and other animals use similar ways to try to attract a mate. Write down two examples.
3. There is a link between the number of offspring a species produces and the amount of parental care.
   Describe this link.
   (Hint: you may need to look back at your work in Year 7.)
4. In your group, discuss how being at the top of a pecking order helps a bird to survive and breed.
5. Animals fight for mates and territories, but not usually 'to the death'.
   Explain the survival value of this.
6. Write down three examples of innate behaviours of worker bees.
7. Write down three behaviour patterns that worker bees have to learn.

9A.5

1 Look at the pictures of Farmer Kirkby's plants.
Write down:
  a the useful characteristics of the tall plant;
  b the useful characteristics of the small plant.

2 Farmer Kirkby did not get the type of plant that she wanted straight away.
What other combinations of characteristics did her first set of plants have?

3 Which of this first set of plants did Farmer Kirkby breed from, and why?

4 Explain why plant breeders have to make sure that flowers are pollinated only with pollen from the plant that they have selected.

5 Explain why plant breeders:
  a remove the anthers before they burst;
  b cover the flower with a pollen-proof bag;
  c use a paintbrush to transfer pollen from the chosen plant to a mature stigma.

9A.6

1 What is a clone?

2 Write down two differences between asexual reproduction and sexual reproduction.

3 Describe the difference in the amount of variation resulting from sexual and asexual reproduction.
Explain your answer.

4 Which part of the rhubarb plant can you use to make cuttings?

5 Explain why taking cuttings is useful for a farmer who makes a living by growing rhubarb.

6 How soon would you expect the farmer to be able to harvest rhubarb from the new plants?
Explain your answer.

7 A farmer grows each new rhubarb plant in the same conditions.
Explain why all the plants will be identical.

8 Look at the diagram of plant grafting.
Why are the two surfaces taped together?

9 Explain the advantages of growing clones of apple trees.

10 Lots of people are against cloning animals such as sheep.
In your group, discuss some possible reasons for this.

9A.HSW

1 We change plants and animals by selective breeding.
In your group, think of as many different examples of selective breeding as you can in 30 seconds.

2 In your group, discuss these two opinions about the ethics of spending money on seed banks.

- 'It is unethical because more people would benefit if we spent the money on getting rid of diseases such as malaria.'
- 'We have a moral responsibility to preserve biodiversity, to prevent the loss of any more species and of variability within species.'

3 We have to store seeds at low temperature and moisture content, so it is expensive to store them.
In your group, discuss who should pay for seed banks and why.

4 In your group, discuss which of the two seed banks described here is likely to:

a cost more in energy;

b cost more to staff.

5 In your group, think of some reasons why some farmers keep rare breeds of cattle, pigs, sheep and chickens.

6 **a** In a small group, discuss:

- the rights and wrongs of breed standards for dogs that lead to health problems;
- whether some breed standards should be changed.

b Make a poster to show possible opinions for and against breed standards.

7 Look at the three examples of genetic engineering.

a Vote on each one to say whether or not you think that it is ethical.

b Discuss your ideas and reasons, then vote again.

c If anyone changed their mind, find out why.

8 You have given your opinions about seed banks, selective breeding and genetic engineering.
Discuss how your opinions might differ from those of people living in the developing world.

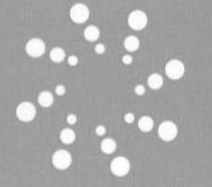

Scientific enquiry

9B.1 Ideas about fitness

You should already know

Outcomes

Keywords

We often hear statements such as 'She's super-fit!', 'I really need to get fit', 'I've never been fitter!'. But we need to think about what it means to be fit.

A

B

C

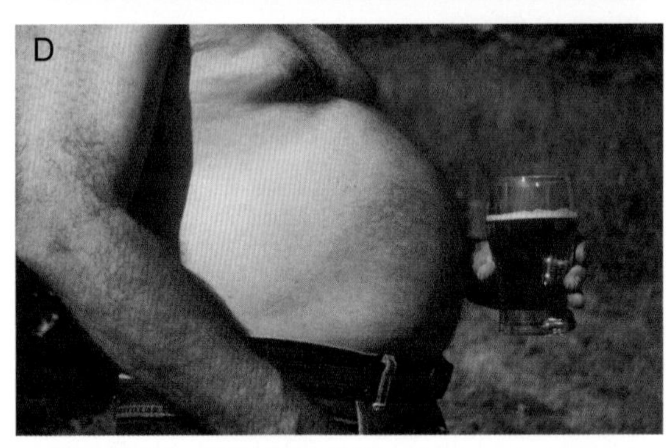
D

E

F

Question 1

Fitness means different things to different people. Just because they're disabled or old doesn't mean that a person is unfit. The fat young man in the photograph is probably very unfit for his age. If you are fit, your body uses energy in an efficient way during exercise. Usually, the fitter you are, the more exercise you can do without getting tired.

Question 2

Energy gets you going

The cells in your body require supplies of glucose and oxygen to release the energy they need.

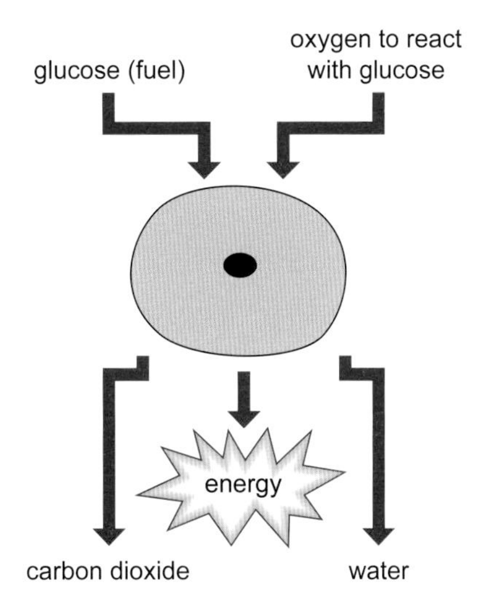

Cells release energy in a process called **respiration**.

Question 3

Different systems in your body work together to keep you fit.

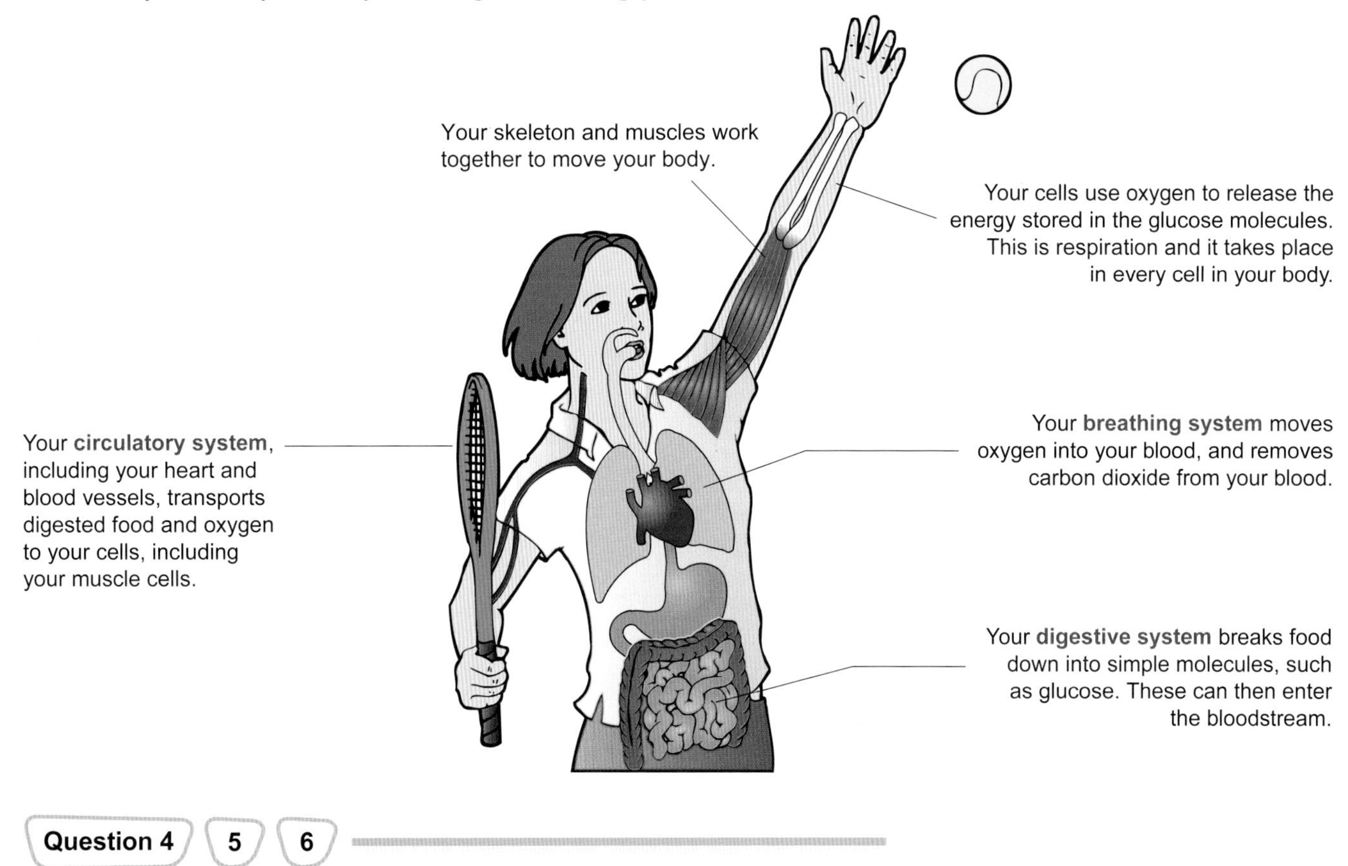

Question 4 **5** **6**

9B.2 Breathing in action

You should already know | Outcomes | Keywords

Working well!

You have learned that every cell in your body takes in oxygen for respiration and releases carbon dioxide. The air containing these gases goes into and out of your lungs when you **breathe**. Your chest works a bit like a pair of bellows, drawing the air in and then forcing it out.

Place one hand on the middle of your upper chest. Place your other hand on your side, just above your waist. Take a few deep breaths.

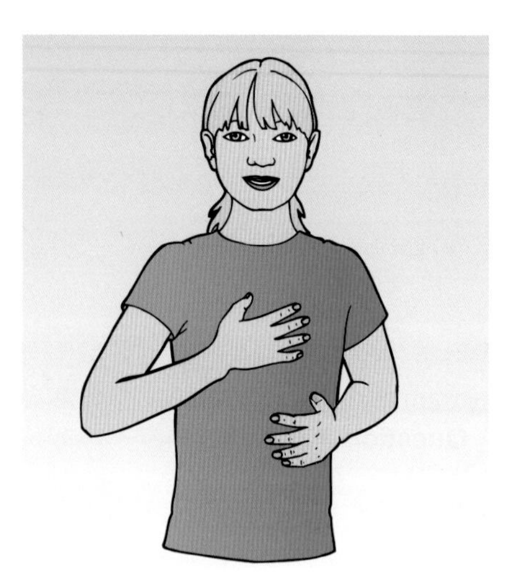

Feeling breathing movements.

Question 1 | 2

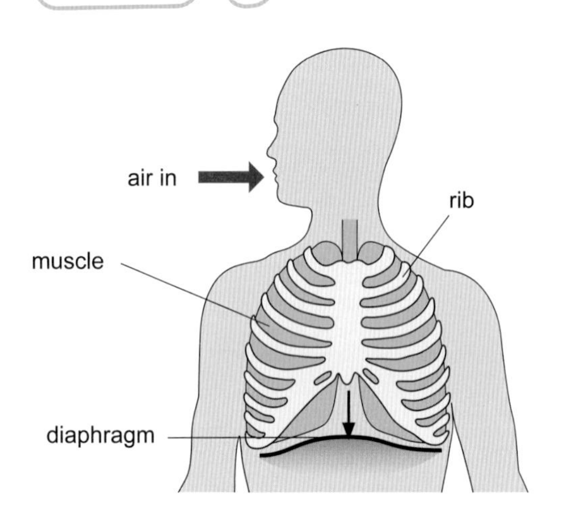

Breathing in.

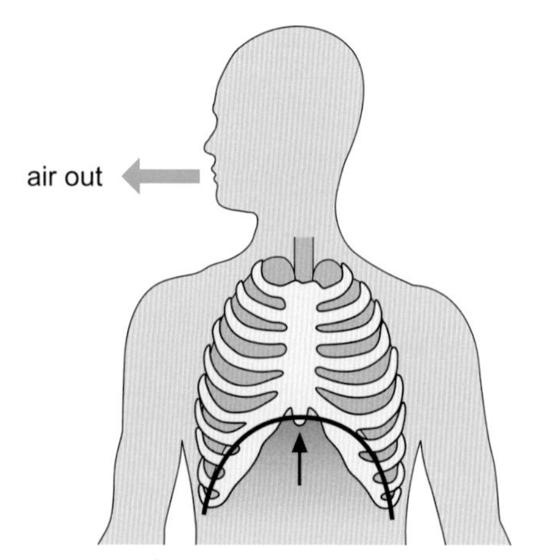

Breathing out.

- The diaphragm is a sheet of muscle that separates the chest cavity from the abdomen. When it contracts, it moves down and becomes flat. This increases the space inside the chest cavity.
- The muscles between the ribs contract, making the ribs move up and out. This increases the volume of the chest cavity. So it lowers the air pressure in the lungs.
- Air moves into the lungs.

- The diaphragm relaxes. It moves up and becomes dome-shaped, decreasing the volume of the chest cavity.
- The muscles between the ribs relax, so the ribs move down and in. This decreases the volume of the chest cavity and increases the air pressure in the lungs.
- Air moves out of the lungs.

Question 3  4  5  6

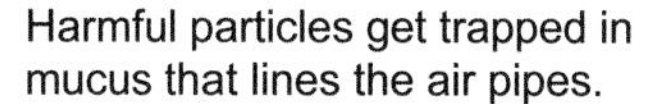

Working badly

In a healthy person, the cilia and mucus in the air pipes work together to keep harmful particles out of the lungs. But sometimes the cilia can't cope.

- There may be too many particles.
- There may be too much mucus.
- Poisonous substances in cigarette smoke may have destroyed the cilia.

Instead of getting swept out of the lungs, the mucus and particles trickle down into them.

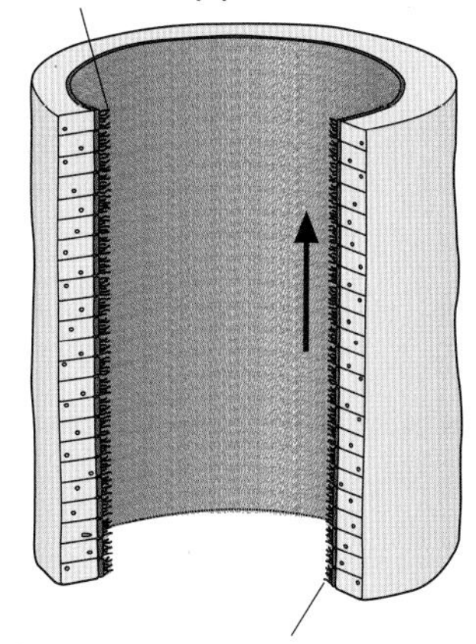

The lining of the air pipes is designed to keep harmful substances in the air, such as dust, bacteria and viruses, out of the lungs.

Question 7

When the person coughs to try and get rid of the mucus, bacteria may remain and cause an infection called **bronchitis**.

Constant coughing can break the walls of the air sacs. This illness is called **emphysema**. It reduces the surface area of the lungs, so the person struggles to get enough oxygen. We say that he or she is short of breath.

Question 8 **9**

The tar in cigarette smoke can make the normal cells lining the air tubes change into cancer cells. These cells divide more than normal cells, and they form a lump called a tumour.

Lung cancer is difficult to cure.

70 per cent of the tar breathed in with cigarette smoke stays in the lungs. It forms a black coating inside the tiny tubes and air sacs.

> **Remember**
> We now group together the lung diseases that cause shortage of breath.
> We call them chronic obstructive pulmonary disease (COPD).

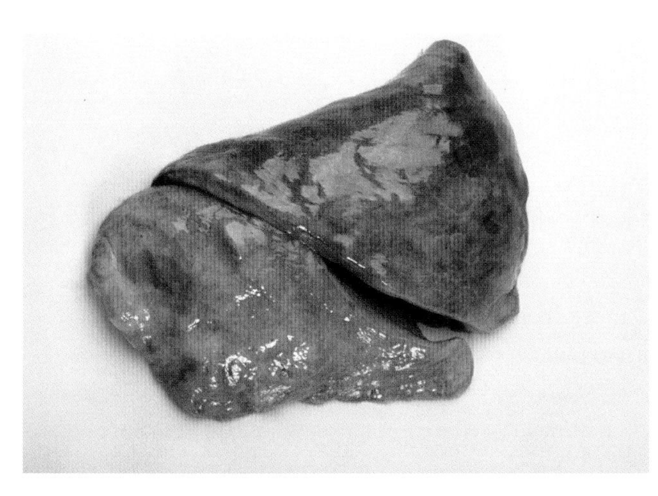

This is what a healthy lung looks like.

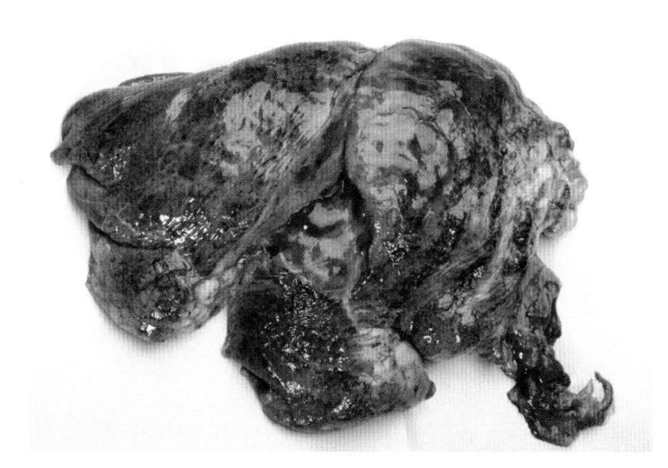

This lung from a smoker is blackened with tar.

Question 10 **11**

9B.3 More dangers of smoking

You should already know | **Outcomes** | **Keywords**

Every year, more than 100 000 people die in Britain as a result of smoking. This is because cigarette smoke contains many harmful substances.

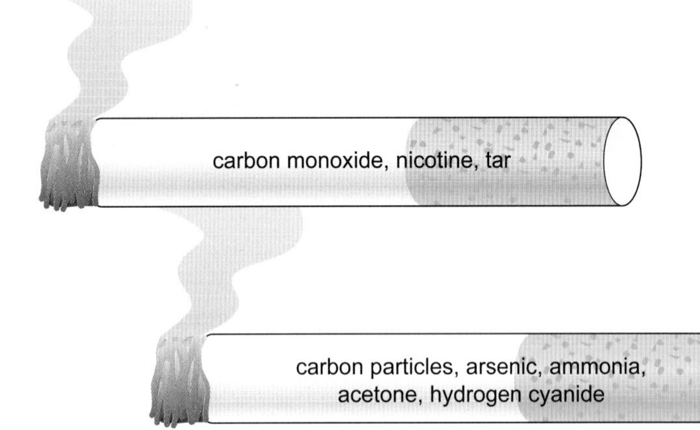

Fact file

- Cigarette smoke contains more than 4000 chemicals, many of them harmful.
- The longer you smoke, the greater your risk of developing:
  - cancer somewhere in your body;
  - COPD.
- Women who smoke during pregnancy are more at risk of:
  - miscarriage;
  - having very small babies.
- Smoking can cause eye and ear problems.
- Smoking increases:
  - tooth decay;
  - weakening of bones (osteoporosis).
- Treating illnesses related to smoking costs the National Health Service (NHS) about £1.7 billion a year.

Question 1

Carbon monoxide is the same poisonous gas that comes out of a car's exhaust pipe. It replaces oxygen in your blood so important organs such as your heart and brain don't get enough oxygen.

Nicotine is an **addictive drug** that makes your body want more of it. When people's bodies become dependent on a drug, we say that they are addicted. Nicotine is as addictive as cocaine.

Nicotine also raises the blood pressure and the heart rate, and makes the heart work harder.

So, the carbon monoxide and nicotine together make smokers more likely to have heart attacks and strokes. A stroke is when a clot of blood causes brain damage or death.

Tar is a mixture of chemicals that cause **cancer** of the mouth, throat and lungs, as well as other lung conditions.

Question 2

Look at the graph.

Question 3 4 5

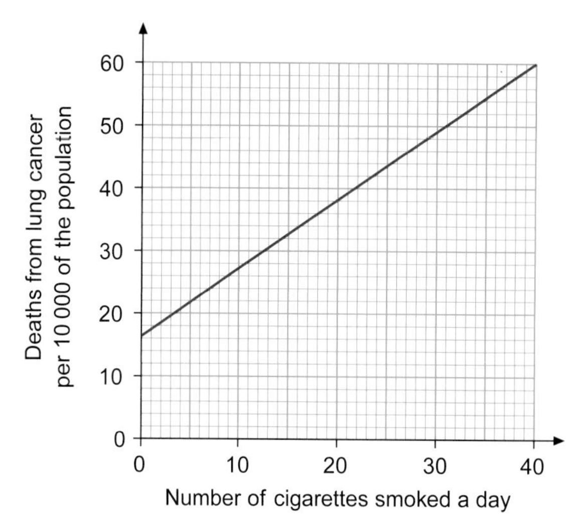

The number of deaths related to smoking in England and Wales.

Remember, this graph doesn't mean that if you smoke you will get lung cancer, or that if you don't smoke you can't get lung cancer. It means that your **risk** of getting lung cancer increases as the number of cigarettes you smoke increases.

Smoking increases your risk of:

- many other unpleasant illnesses;
- becoming disabled;
- dying young.

Some things about being a smoker also make you less attractive – look at the diagram.

Most adult smokers started smoking in their teens. About half will die of an illness related to smoking.

Even by the age of 20, smokers:

- are less **fit** than non-smokers;
- are likely to have tiny wrinkles around their mouths.

By the age of 30, the average smoker will have spent £28 000 on cigarettes. Smokers probably:

- look older than they are;
- have stained teeth and fingers and dull, smelly hair;
- will be less fertile and have babies more at risk of cot death.

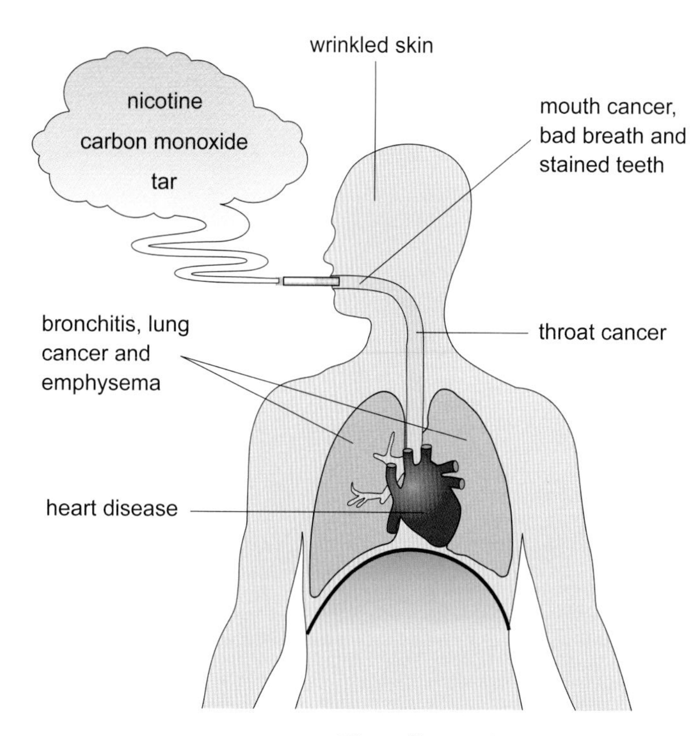

The effects of smoking are not 'cool' or exciting.

Question 6 7

The good news

The damage caused by smoking need not be permanent. If a heavy smoker stops smoking, his or her risk of getting lung cancer slowly falls. Within 10 years, that chance is the same as for a person who doesn't smoke. So, it's never too late to STOP!

Question 8

Check your progress

9B.4 Why your diet is important

You should already know | Outcomes | Keywords

A balancing act

To be fit and healthy, it is important to eat a **balanced diet**. This is one that contains the right amounts of proteins, carbohydrates, fats, vitamins and minerals, fibre and water.

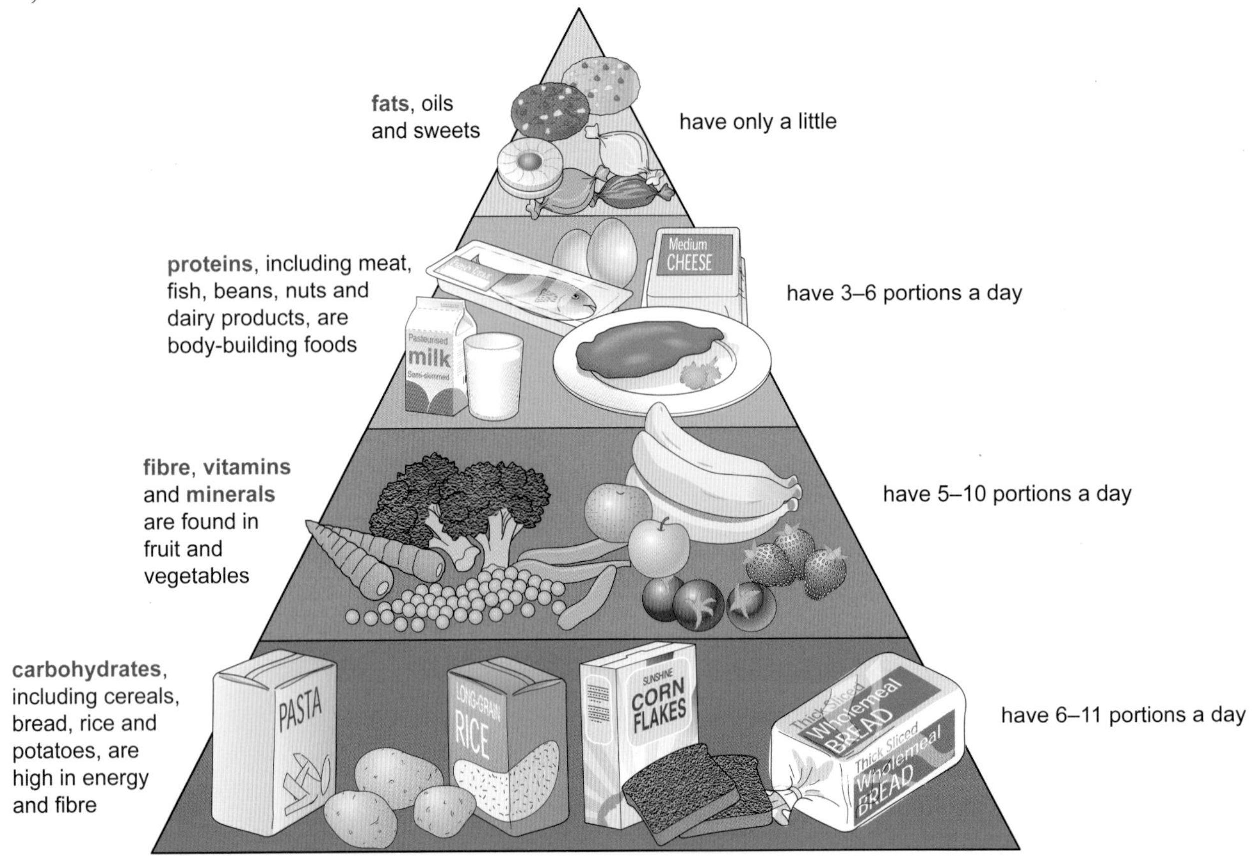

A pyramid is an easy way of showing how much food from each group you need to eat.

Question 1 | 2 | 3

For centuries, scientists have been gathering evidence about how what we eat affects our health. For example, in the early 1900s a famous English scientist, Frederick Gowland Hopkins, fed rats on a diet of pure carbohydrates, fats, proteins and minerals. He thought that these were all the things needed for good health. Within weeks, all the rats were dead. Another group of rats had the same diet but with a little milk added, and they all survived. Something in milk, needed in tiny amounts, kept the rats healthy. We now call these things vitamins.

Question 4

Feast or famine?

Malnutrition means 'bad eating'. A malnourished person eats too much or too little of a particular food. In Britain, most malnourished people either eat the wrong foods or they eat too much food. Food is your body's source of energy. If you eat more food than you need, the extra is stored as fat and you can become overweight or obese.

To lose weight, you need to eat less and exercise more to use the stored fat.

In some countries, many people die of starvation. Others suffer from kwashiorkor, which is malnutrition caused by a lack of protein. Many people worldwide have other diseases caused by a shortage of just one thing in their diet. We call these **deficiency diseases**.

Nutrient	Source	Symptoms of deficiency
iron	red meat eggs cereals	anaemia: people look pale and feel weak; their blood does not carry enough oxygen
calcium	dairy products dark green vegetables sardines (bones)	poor growth of bones and rotting teeth
vitamin C	citrus fruit fresh vegetables	scurvy: gums crack and wounds don't heal; bleeding under the skin
vitamin D	eggs dairy products fish oil made by your skin in sunlight	rickets: softening and bending of the bones
protein	meat fish eggs dairy products pulses (peas, beans, lentils)	kwashiorkor: poor growth, swollen bellies because water collects in the tissues

A deficient diet can lead to many diseases or problems.

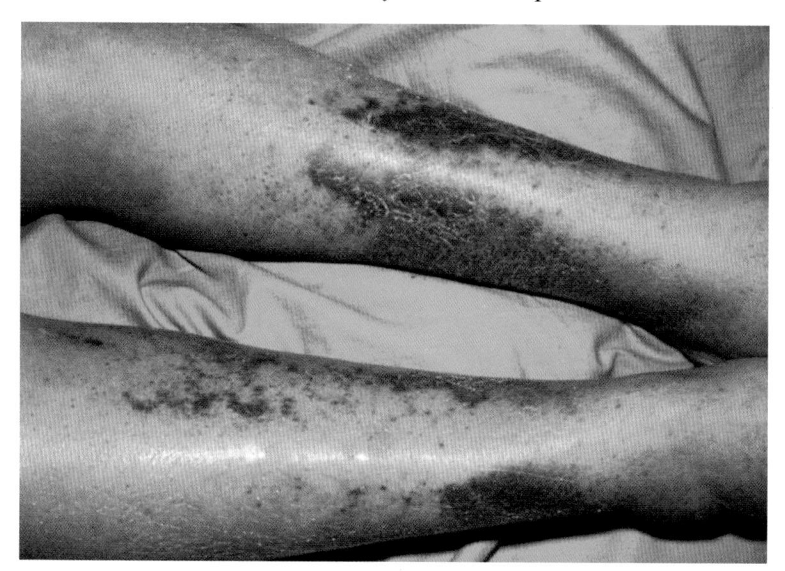

This person has scurvy.
In this disease, blood vessels below the skin burst.

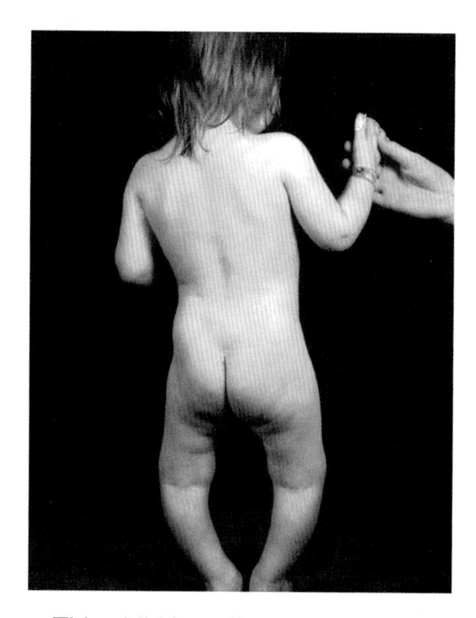

This child is suffering from rickets.

Question 5

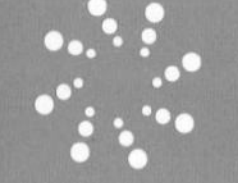

9B.5 The use and abuse of drugs

You should already know

Outcomes

Keywords

A **drug** is any substance that changes the way your body or mind works. Some changes are helpful, others harmful. Some drugs improve your mood, feelings or behaviour; others make them worse. Some drugs heal, and others damage organs in your body. Even drugs that heal can have bad effects, such as headaches and sickness. We call these **side effects**.

This diagram shows just one of the ways of grouping drugs.

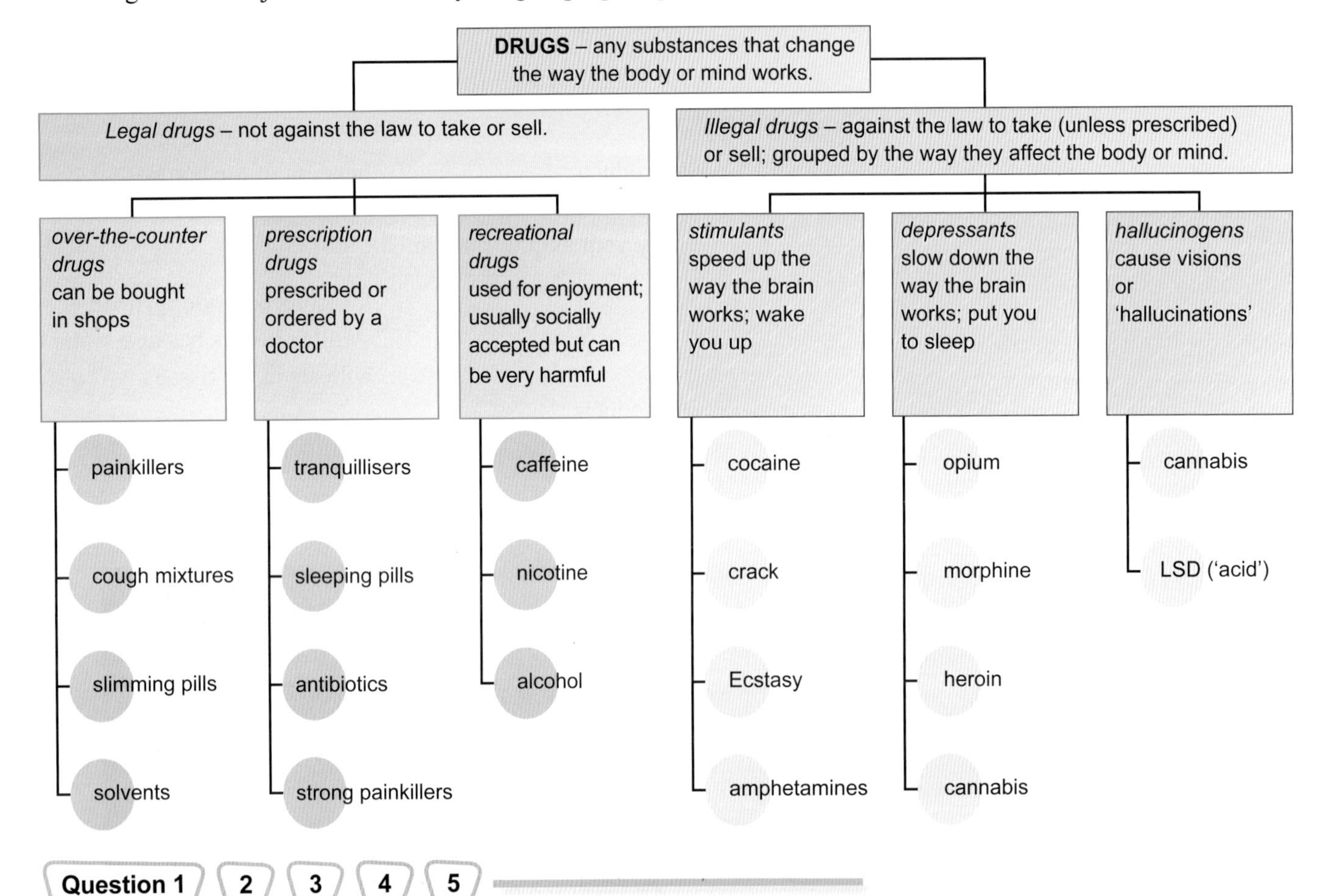

Question 1 2 3 4 5

<u>All drugs can be dangerous</u>. When you use a legal drug, you should follow the instructions very carefully. If you don't, or if you use a drug prescribed for someone else, then you are misusing or abusing the drug. Abuse of both legal and illegal drugs is dangerous.

<u>Many drugs are addictive</u>. An **addict** is dependent on a drug and needs it to function. An addict who stops taking a drug gets **withdrawal symptoms** – feels ill and anxious, may shake or vomit.

Question 6 7

Some effects of alcohol

Alcohol is a drug that affects both mind and body. It is also a poison. Your liver helps to protect you by breaking alcohol down into less harmful substances. It takes about one hour for a healthy liver to break down one unit of alcohol. Your liver cells become damaged if you drink alcohol at a faster rate than this, or if you drink heavily over a long period of time. This damage is called cirrhosis and can cause death. Health advisers suggest that children shouldn't drink and that adults should drink no more than 14 units a week (women) and 21 units (men).

Alcohol slows down your reactions and affects your behaviour.

- It can reduce your will power and self-control. This can lead to violence, accidents and unwanted pregnancies.
- It can make you sleepy and clumsy and your speech slurred.
- It can cause loss of memory and even permanent brain damage.
- It can increase your blood pressure and pulse rate. Over a long period, alcohol increases your risk of a heart attack.
- It can irritate the lining of the stomach and can lead to stomach ulcers.
- It can increase the flow of blood to the skin. This makes you look flushed and feel warm. But this makes the body loses heat quickly. So, drinking alcohol in cold conditions can be dangerous.
- It can pass to a developing baby through the placenta. This can slow the baby's development and even cause brain damage.

Question 8 **9**

Alcohol is legal and used regularly by many people. But remember:

- Twice as many people are addicted to alcohol as to all other drugs put together.
- Violent crime and violence in the home are often linked to alcohol.
- More than 1000 young people are admitted to hospital each year with alcohol poisoning.
- If you drink heavily in the evening, you can still be over the legal drink-drive limit the next morning.

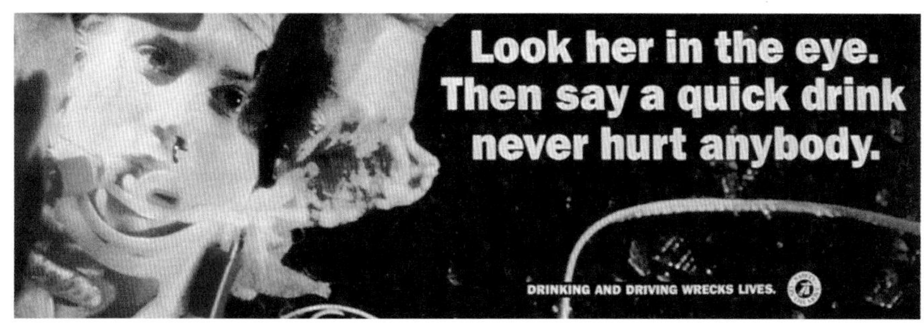

About half of adult pedestrians killed in road accidents have had too much to drink. So, think before you drink!

Question 10 **11**

As a rough guide:

1 unit = half a pint of beer

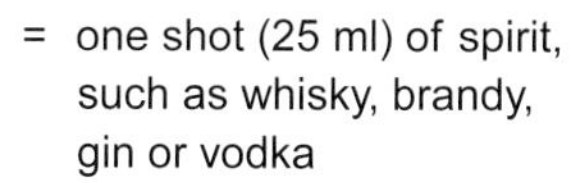

= one shot (25 ml) of spirit, such as whisky, brandy, gin or vodka

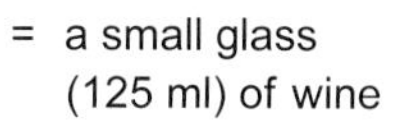

= a small glass (125 ml) of wine

= 150 ml of 'alco pop'

One way of comparing the amount of alcohol in different drinks is by using 'units' of alcohol as a measurement. A unit is 10 ml of pure alcohol.

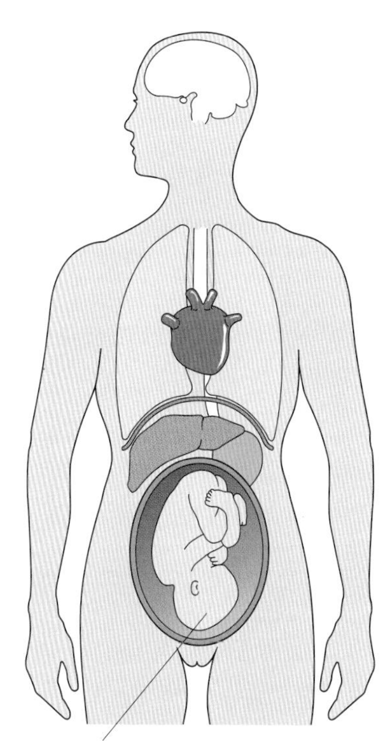

Alcohol can cause brain damage and stunt the growth of a developing fetus.

Some parts of the body affected by alcohol.

9B.6 Fit for life

You should already know | Outcomes | Keywords

So far, in this unit, you have looked at the effects of diet, drugs and exercise on health and fitness.

Exercise is important for keeping you fit and healthy. However, too much exercise, or the wrong kind of exercise, can damage joints and muscles.

A **joint** is the place where two bones connect. **Muscles** are fixed to bones on both sides of a joint. The muscles pull on the bones to make you move.

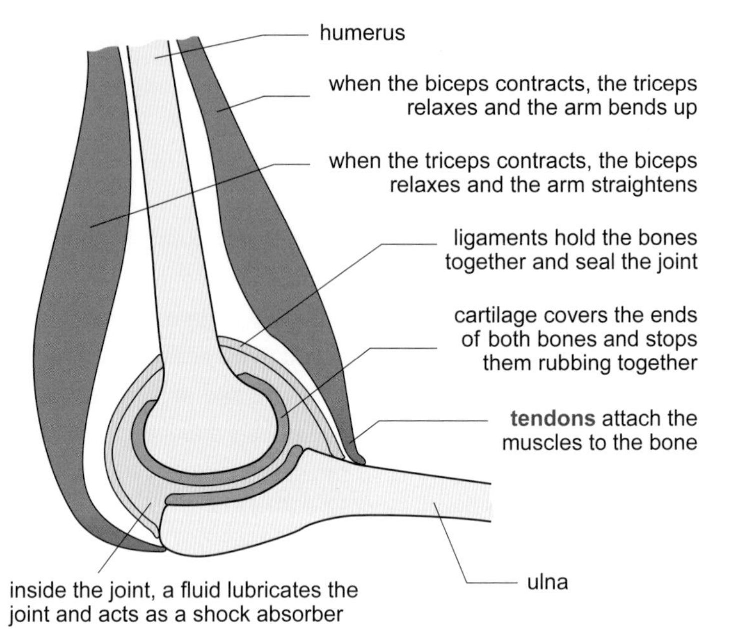

An elbow joint.

Question 1 | 2

If you twist your foot inwards suddenly, you can overstretch or tear the **ligaments** of the joint. Your ankle swells and hurts. This is a sprain.

If you do sudden vigorous exercise, you might injure a muscle. This is called a strain or 'pulled muscle'. Overdoing exercise can sometimes make muscles contract so powerfully that they hurt. This is called cramp.

Question 3 | 4 | 5 | 6

Exercising for health

If you exercise regularly, you can improve your fitness in different ways.

- Your heart is mainly muscle, and exercise makes it stronger and more efficient.
- Exercise strengthens your muscles. This helps to prevent injury to muscles and joints.
- Exercise develops the blood vessels in muscle. More oxygen and carbon dioxide can flow to and from the muscles, so your muscles work better.

You can reduce the risk of getting a sports injury if you warm up before exercise.

Eating for health

If you eat a diet high in fat, you increase your risk of **coronary heart disease**.

To lower your chance of heart disease, you need to:

- eat at least five portions of fruit and vegetables a day;
- eat a low-fat diet;
- keep your weight normal, so your heart doesn't need to overwork;
- not eat too much salt, because salt raises your blood pressure.

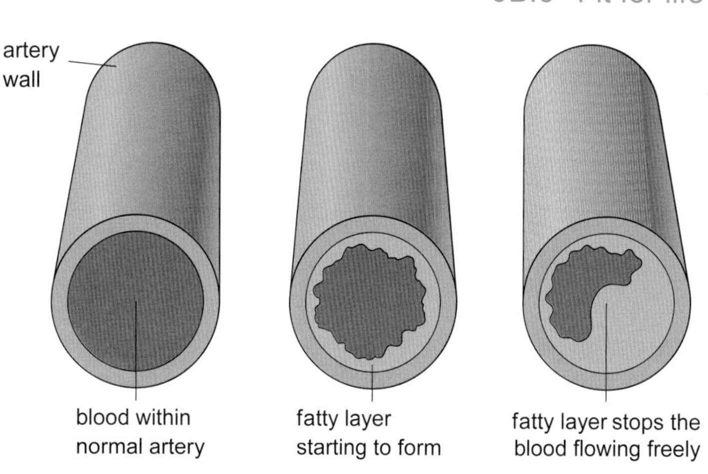

Coronary heart disease begins when fatty material builds up inside arteries in the heart. This makes the space inside smaller, so your heart has to work harder to push the blood through. The heart muscle becomes short of oxygen, and this causes **angina**, which is very painful. A heart attack happens when a blood clot blocks the narrowed arteries. The fatty material is made from cholesterol, which the body makes from fat in food.

Question 7 | 8

Are we healthier than our great-grandparents were?

Science is about finding answers to questions. The question in the heading is a general question. Scientists first break down a general question into a few smaller questions that are easier to investigate.

They then plan a suitable investigation and collect the evidence.

They use the evidence to answer the smaller questions. Lastly, they use these answers to help them answer the general question.

Your first response to the question 'Are we healthier than our great-grandparents were?' might be 'Yes, of course we are'. But you need to investigate this general question in a scientific way – finding evidence to back up your answer.

First, you need to think about differences between your lifestyle and that of your great-grandparents that might have an effect on health – different diets, for example. Then you think of questions to investigate.

Question 9 | 10

9B.7 Psychology and behaviour (HSW)

You should already know

Outcomes

Keywords

Ethology is the science of observing behaviour in natural surroundings.

Psychology is the study of what people (and animals) do and why they do it. Like other scientists, psychologists ask questions. Then they use observation and experiment to try to answer the questions. They study:

- intelligence and learning
- moral and social development and relationships
- self-harming and anti-social behaviour.

They don't just explain people's behaviour. They try to help people to change behaviours that cause problems.

> **Information**
>
> Some branches of psychology:
>
> Behavioural – understanding thought, feelings and behaviour.
>
> Social – how people interact.
>
> Cognitive – attention, memory and learning, problem solving, etc.
>
> Applied – includes help with mental health, learning and relationships.

Questions psychologists ask about behaviour

A Why do some people need help with problems caused by sex hormones?

B Do their genes make some people more at risk of becoming addicts than others?

C How can we help people with anorexia (who are very thin) to stop seeing themselves as fat?

I lose control and hit people. So I've been sent on an anger management course.

D This is called the placebo effect. Does it work for all types of people?

E Why do some people find it particularly hard to control their temper?

F What makes some people take risks that put them and others in danger?

Question 1 **2** **3**

Investigations in psychology

Laboratory investigations are designed to be objective and repeatable by others. That is not always possible in investigations of humans.

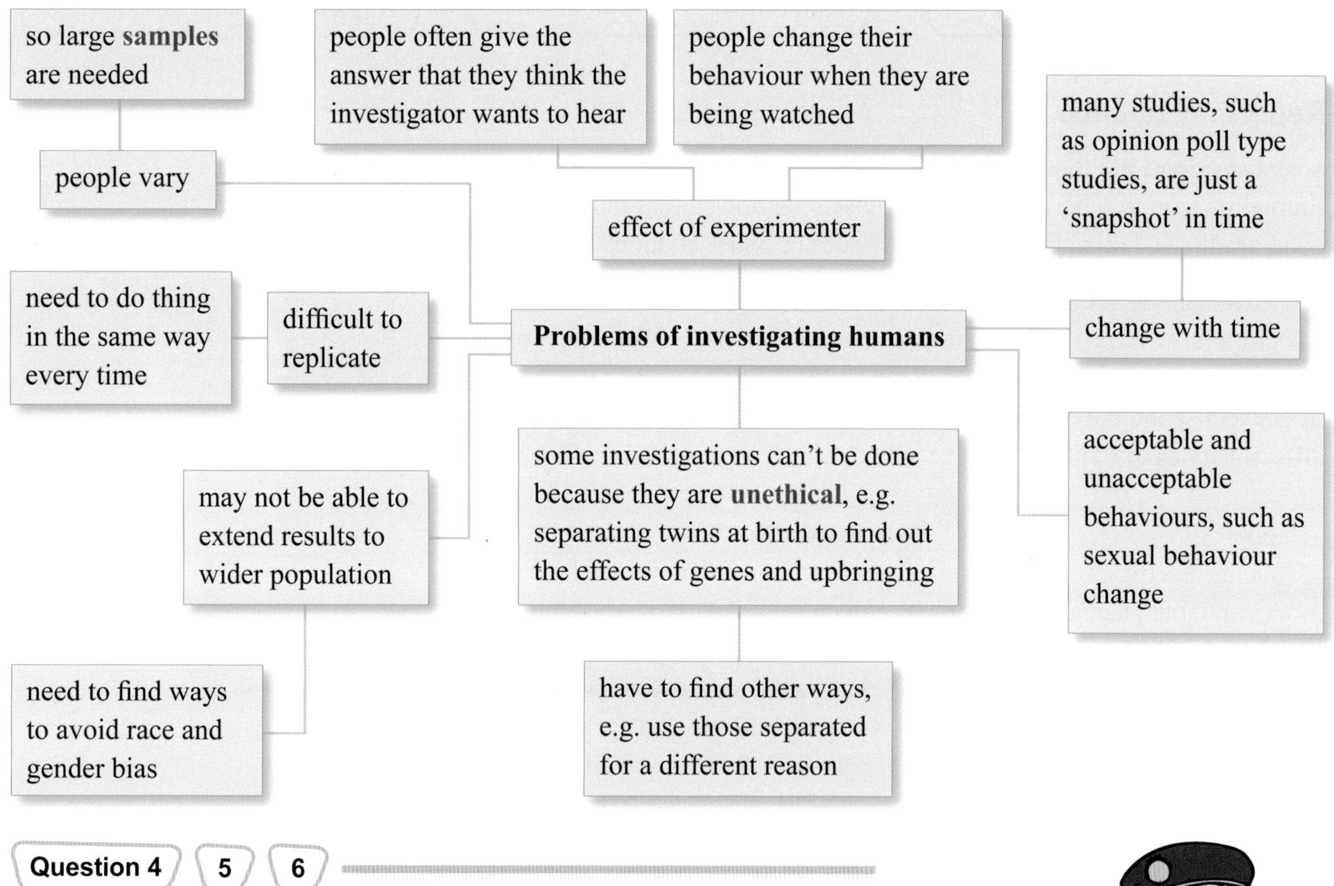

Question 4 5 6

In 1974, Stanley Milgram did an experiment that raises some ethical issues. Milgram investigated the conflict between 'following orders' and personal conscience. He asked the people who were the subjects of his experiment:

- to act as 'teachers' to help him to study the effects of punishment on learning behaviour;
- to give the learner an electric shock for each mistake made – each shock worse than the one before.

Shocking people made some 'teachers' stressed and uncomfortable. He encouraged them to continue. They didn't know that the 'learner' was an actor and was not actually given electric shocks.

Look at the table.

Some war criminals use this excuse.

Intensity of electric shock	'Teachers' continuing to give shocks
300 V	100%
450 V	60%

(The mains voltage in the UK is 230 V.)

Question 7 8 9

Review your work

Summary

9B.HSW Health in the media

You should already know

Outcomes

Keywords

Reporting science

Scientists report the results of their investigations in **scientific journals**. Journalists tend to report in the media those scientific studies that make good headlines. Often, these are health issues or issues about behaviour.

You have learned about some of the benefits of exercise and about some risks of injury. Surprisingly, exercising too hard can have a bad effect on your **immune system**. Read the two reports of the same study. One is by a scientist and the other by a journalist. Journalists and scientists use different styles of writing.

Report A
Moderate exercise is beneficial for health, but acute and vigorous exercise can be harmful. The data show that the physiology and viability of circulating leucocytes can be changed by over-strenuous exercise causing immune distress.

Report B
We all know that we need to exercise regularly if we want to stay healthy. But a recent report shows that overdoing it can make you ill. You can end up with fewer white cells in your blood, and this makes you less able to fight disease.

Question 1 2

The researchers asked 12 healthy athletes to exercise hard for 30 minutes a day for 3 days.

They took blood samples immediately before and after each exercise session. Then they took further samples 24 and 72 hours after the final session.

Blood test	At the end of exercise	24 hours later	72 hours later
inflammation markers in the blood	raised	raised	raised
MTP markers in white blood cells (these markers show how well the cells are working)	lowered	lowered	normal
death rate of white blood cells	increased	increased	increased

Question 3  

White blood cell counts of football players

Another study followed the change in the numbers of one kind of white blood cell in Manchester United football players. From July to November the numbers remained normal. By the following May, the numbers had dropped to less than one third of normal.

Question 6 **7**

A newspaper headline can grab your attention

The headline is for a report of research carried out by the University of Cambridge and the Medical Research Council between 1993 and 2006.

Healthy living 'can add 14 years to your life'

A textbook report of the research

The researchers surveyed 20 000 people in Norfolk aged 45–79. The people were a socially mixed group and seemed to be free of **cancer** and **heart disease** at the start of the study. The table shows how the researchers scored **lifestyle factors**.

Healthy lifestyle factor	Score
non-smoker	1
no more than 14 units of alcohol per week	1
5 servings of fruit and vegetables per day	1
active lifestyle	1

Question 8

Professor Kay-Tee Khaw led the research.

He said 'We've known that, individually, measures such as not smoking and exercising can have an impact upon longevity, but this is the first time we have looked at them all together'.

Question 9 **10** **11**

The results

Total score	% alive in 2006 (approx.)
0	76
1	85
2	89
3	92
4	95

Those with 0 points were more likely to have died over the period of the study than those with 4 points.

How overweight or poor people were did not affect the findings.

A 60-year-old person with a score of 0 had the same risk of dying as a 74-year-old with 4 points.

The results suggest that many people could live for longer if they made some lifestyle changes.

Question 12 **13**  **14**  **15**

9B Questions

9B.1

1 Look at the photographs. Then discuss these questions in a small group. Give reasons for your decisions.
   - **a** For each person, decide whether or not you think they are fit.
   - **b** Which person shown is probably the most unfit?

2 In a small group, try to agree on a list of different kinds of fitness.

3 Write a word equation for respiration.

4 Why do all your cells need glucose and oxygen?

5 Look at the diagram.
Write down three systems in your body that help to get glucose and oxygen to your cells.

6 Choose one of the three systems and explain briefly how your fitness is affected when that system is not working properly.

9B.2

1 Describe the movements that you feel, first as you inhale (breathe in), and then as you exhale (breathe out).

2 To model breathing, cut out an A shape from a piece of paper about 8 cm square. Move the 'legs' of the A in and out.
What happens to the 'cross-bar' and the volume inside the A?

3 Name four parts of your body that are involved in breathing.

4 Explain what makes air go into your lungs.

5 Which requires more effort, breathing in or breathing out?
Explain why.

6 What happens to your breathing rate when you:
   - **a** exercise?
   - **b** go to sleep?

7 How does the breathing system work to stop harmful substances in air getting into the lungs?

8 Explain why smokers often cough when they wake up in the morning.

9 Find out more about either bronchitis or emphysema. In your research, include the symptoms of the illness and how it is treated.

continued

10 In a small group, write down as many harmful effects of smoking as you can in one minute.

11 Make a combined list for the whole class.
Draw a star next to any idea on your list that no other group had on their list.

9B.3

1 How many different chemicals are there in cigarette smoke?

2 Draw a table to show the dangerous substances in cigarette smoke. In one column, name the substances; in the other column, list the parts of the body that are most affected by them.

3 What does the graph show?

4 How many people (per 10 000) who don't smoke die of lung cancer?

5 How many people (per 10 000) who smoke 20 cigarettes a day die of lung cancer?

6 Write down three ways that smoking can make you less attractive to other people.

7 Smoking harms health and costs the National Health Service (NHS) a lot of money.
Discuss in your group:

a Should smoking be banned completely?

b Should the NHS pay for a heart operation needed by a smoker?

Remember that when you discuss issues such as these, you should consider the arguments for and against and have reasons for any decisions you make. People have different opinions and you should respect them.

8 Write down in one sentence what you believe is the most important reason never to start smoking.

9B.4

1 Draw a table to summarise the main food groups and examples of foods that contain them.

2 Explain why it is better to get most of your energy from carbohydrates such as cereals, potatoes and pasta, rather than from fats, oils and sweets.

3 Plan a healthy, balanced picnic meal. It should taste good too! Use the pyramid to help you.

4 a Explain how the evidence gathered by Frederick Gowland Hopkins affects your life.

b In your group, argue the cases for and against relying on evidence from experiments on animals in relation to human diet.

5 Explain what you need to add to your diet if:

a your bones and teeth are weak;

b you are suffering from kwashiorkor.

9B.5

1 Name three groups of legal drugs.

2 Are any legal drugs harmful or dangerous? Use examples to explain your answer.

3 Write down one example of a drug that falls into more than one group. Explain your answer.

4 Many drugs have side effects. Use an example to explain what this means.

5 What do we call drugs that can be taken only with a doctor's permission?

6 Name two legal drugs that are addictive.

7 Drugs can be harmful, have side effects and be addictive. Taking addictive drugs can lead to crime. In your group, discuss some other bad effects that taking drugs can have on people's lives.

8 List four organs in the body that can be damaged by alcohol.

9 Read again through the section on the effects of alcohol on the body. Summarise the main points by annotating a diagram of the human body. An example has been done for you.

10 Use examples to explain the difference between the 'use' and 'abuse' of alcohol and other drugs.

11 Some people think that alcohol should be banned or more controlled. Discuss this in your group. Support your views with evidence. Say whether your evidence is based on science or on something else.

9B.6

1. What is a joint?
2. Write a paragraph to explain to a 10-year-old how a joint works.
   Use the diagram to help you.
3. Explain the difference between a sprain and a strain.
4. A first aid manual suggests putting an ice pack on a sprained ankle.
   Suggest why.
5. Find out some of the common causes of sports injuries.
6. Find out about one of the following conditions: tennis elbow, water on the knee, dislocated hip, arthritis, slipped disc or any sports injury.
   Present your findings as a paragraph for a school textbook for 14-year-olds.
7. Explain the cause of angina.
8. You are a doctor seeing a patient who has heart problems.
   List five questions you would ask your patient, to find out if they need to make some lifestyle changes.
9. Write down at least three things affecting health that were different in your great-grandparents' time.
10. Working in a small group, think of one question you could investigate in a scientific way that would help you to answer the general question, 'Are we healthier than our great-grandparents were?'
    Collect together the smaller questions from all the groups in the class.

9B.7

1. Suggest the type of psychologist that would research the questions in boxes A to F.
   There may be more than one answer. Use the information box to help you.
2. Work in small groups. Each group should discuss a different question from A–F and feed back its ideas to the class. Include:
   - some ideas that might help to answer the question;
   - how these ideas could be tested;
   - some practical and ethical issues of these tests.
3. Which type of psychologist might run an anger management course?
4. When investigating humans and other animals, it is important to use a large sample.
   Explain why.

continued

5 When you repeat an investigation, you need to do things in the same way every time.
Explain why.

6 Write down two ways that an investigator can affect the behaviour of humans or other animal subjects of an experiment.

7 What percentage of 'teachers' stopped shocking learners before the shocks reached UK mains voltage?

8 In your group, discuss:

a whether or not you are surprised by the results, giving your reasons;

b what might make people 'follow orders' in this way.

9 Many people say that this experiment was not ethical.
In your group, identify unethical aspects of Milgram's experiment.

9B.HSW

1 In a small group, decide which report, A or B, is from a scientific journal and which is from a newspaper.
Discuss the evidence for your decision.

2 a In a small group, write a headline for the newspaper report.

b Choose the best headline from all the groups.

3 Look at the results table. In your group, agree on some conclusions.

4 This study was limited to 12 subjects and they were all athletes.
Discuss how the researchers should follow up their study. Give reasons for your suggestions.

5 Write a short news report about this research for television. Include suggestions of films or photographs to illustrate it.

6 a Suggest why the number of white cells went down.

b What problems may the players have had as a result?

7 Suggest why journalists reported this study.

8 Suggest why the group chosen for this study was:

a socially mixed;

b 20 000 people.

9 The researchers had to agree on what was an active and what was an inactive lifestyle.
In your group, agree on what you consider is an active lifestyle.

10 What is the score of a person with the healthiest lifestyle?

continued

11 Translate Professor Khaw's comment into a form that a primary school pupil will understand.

12 **a** Draw a graph to show the relationship between lifestyle scores and thc percentage of people who were alive at the end of the study.

b What does the graph show?

13 More people with 0 points than people with 4 points died during the period of the study.
About how many times more?

14 Look at the comments in the speech bubbles.

a Which idea most surprises you and why?

b Add your idea to a tally chart of class results.

15 **a** In a small group, work out a simple way of getting this message about lifestyle across to other people your age.

b Look at the ideas of all the groups. For each one, select:

- the best point;
- one thing that could be improved.

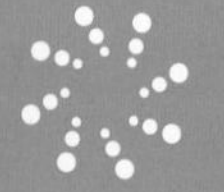

9C.1 How do plants grow? (HSW)

Scientific enquiry

You should already know

Outcomes

Keywords

Food for growth

Only green plants make food. We and other animals eat the food that plants make, or eat other animals that have fed on plants.

We call the mass of all the materials in a plant the **biomass** of the plant.

Question 1

In the 17th century, a Dutch scientist called Jan Baptista van Helmont believed that the biomass of a plant came from just water. He planted a willow tree weighing only 1 kg in a big pot containing 90 kg of dry soil. He covered the soil with a piece of tin with lots of holes in it so that only rainwater went through it. Five years later, the tree weighed 100 kg.

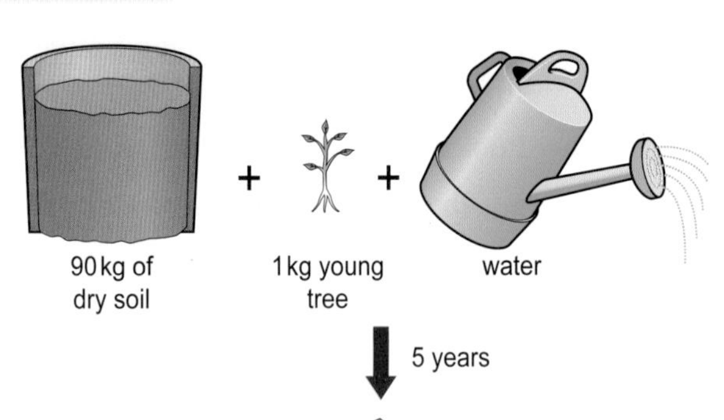

Van Helmont's experiment.

Question 2

Eleanor told her science teacher, Mr Holmes, that she wasn't convinced by van Helmont's experiment. She still believed that plants got their food from the soil.

Mr Holmes set up an experiment that showed Eleanor that van Helmont was right – plants don't need soil, but do need water to make food and grow.

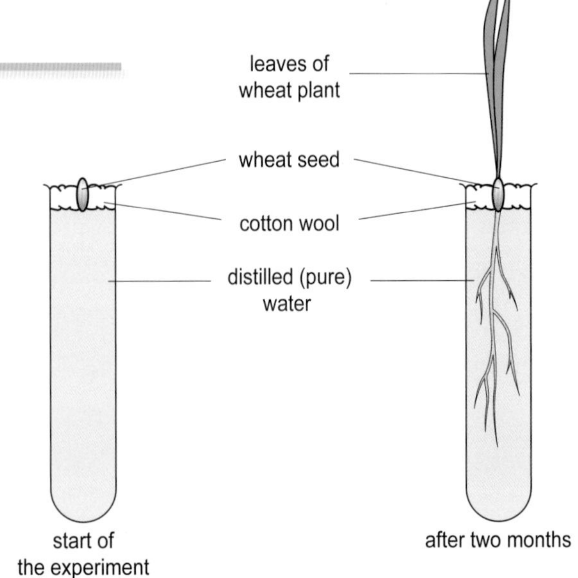

Mr Holmes' experiment.

Question 3

Plants also use a gas from the air

In another experiment, Mr Holmes wanted to find out whether plants took anything from the air. He set up a datalogger to record the carbon dioxide concentration amongst the leaves of a wheat crop.

Time	2 am	5 am	8 am	11 am	2 pm	5 pm	8 pm	11 pm
CO_2 concentration (parts per million) amongst leaves	425	352	300	285	280	287	335	380

Question 4

So, the leaves of plants are like little food factories. The raw materials that they use to make food are:

- **carbon dioxide**, taken from the air through pores in the leaves;
- water, absorbed from the soil by the roots.

The energy for the process of making food comes from light. The word 'photo' means light, and the word 'synthesis' means making, so we call the process **photosynthesis**.

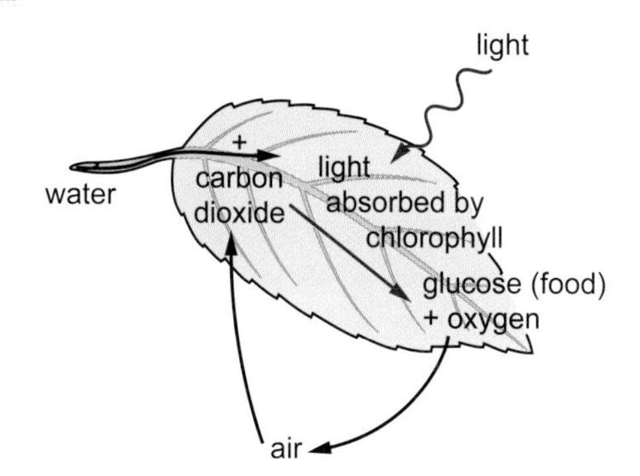

What happens in a leaf.

Question 5

Carbon dioxide in, oxygen out

So, plants take in carbon dioxide from the air. In 1772, Joseph Priestley discovered that plants produce a gas that animals need. We now call this gas **oxygen**.

We can collect the bubbles of gas produced by the water plant *Elodea* and test them for oxygen. If we can collect enough gas, it will relight a glowing splint, so we know it is oxygen.

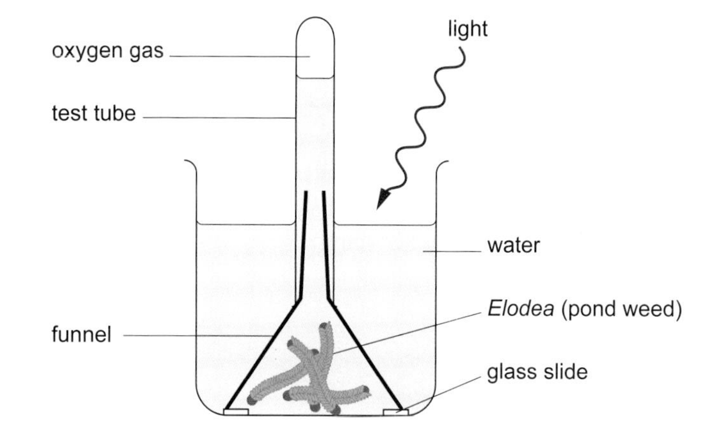

When we do an experiment, we normally set up a **control**. For example, when we look at the effect of water on plants, we need a control experiment in which identical plants are kept in the same conditions, but not given any water.

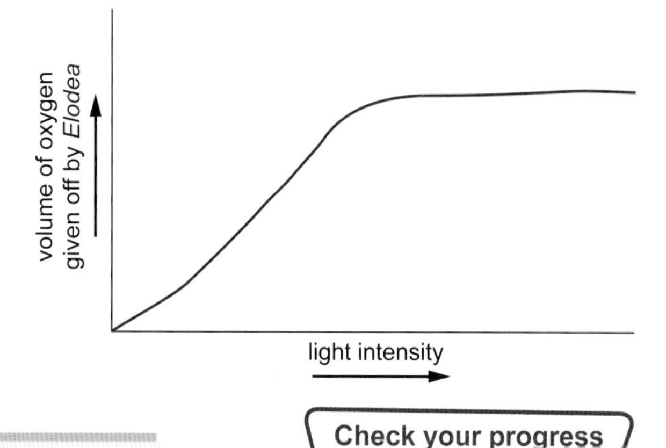

How light intensity affects the amount of oxygen produced.

Question 6  7  8 9

Check your progress

9C.2 Leaves and photosynthesis (HSW)

You should already know

Outcomes

Keywords

Evidence for photosynthesis in leaves

Glucose is a soluble sugar. To keep the concentration constant, leaf cells change any extra glucose into insoluble **starch**. So, we can use the presence of starch as a sign of photosynthesis.

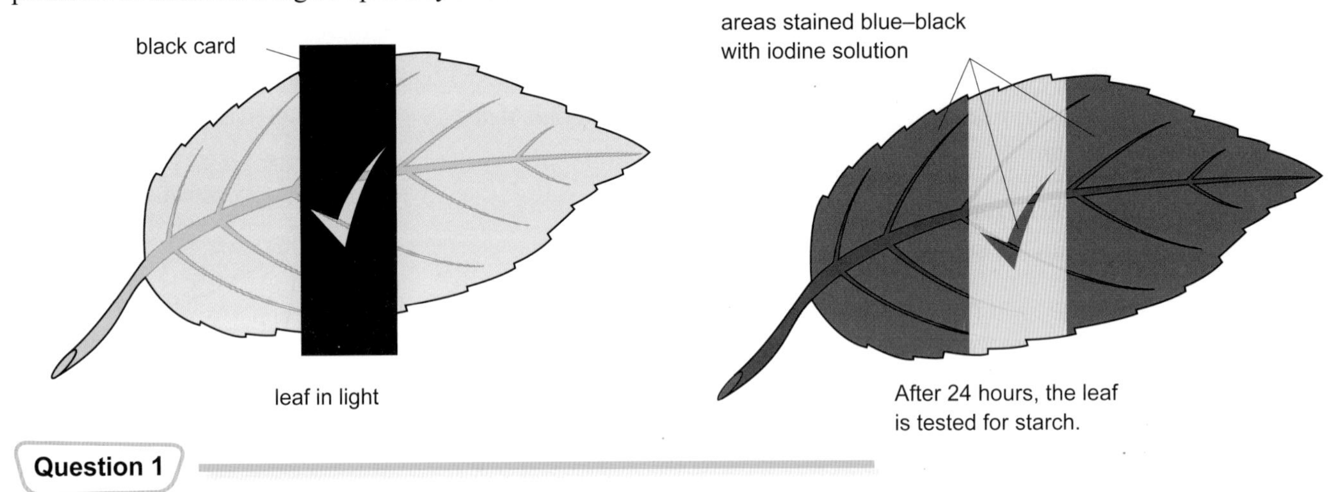

Question 1

The need for chlorophyll

Chlorophyll is the green substance in leaves. Some leaves have green parts (containing chlorophyll) and white parts (with no chlorophyll). We call them variegated leaves. Look at the diagram to see what happens when we test one of these leaves for starch.

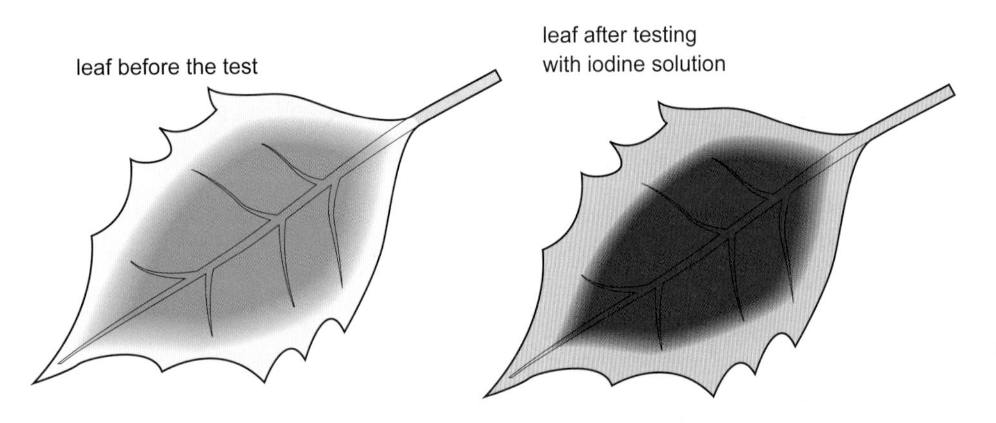

Starch turns blue–black when iodine solution is added to it.

Question 2

Without chlorophyll, a leaf cannot trap and use the light it needs to make food. Plants that are kept in the dark cannot make food. They use food, so they end up weak and spindly.

A closer look at a leaf

Look at the diagram. It shows the cells in which most photosynthesis happens. These are called palisade cells.

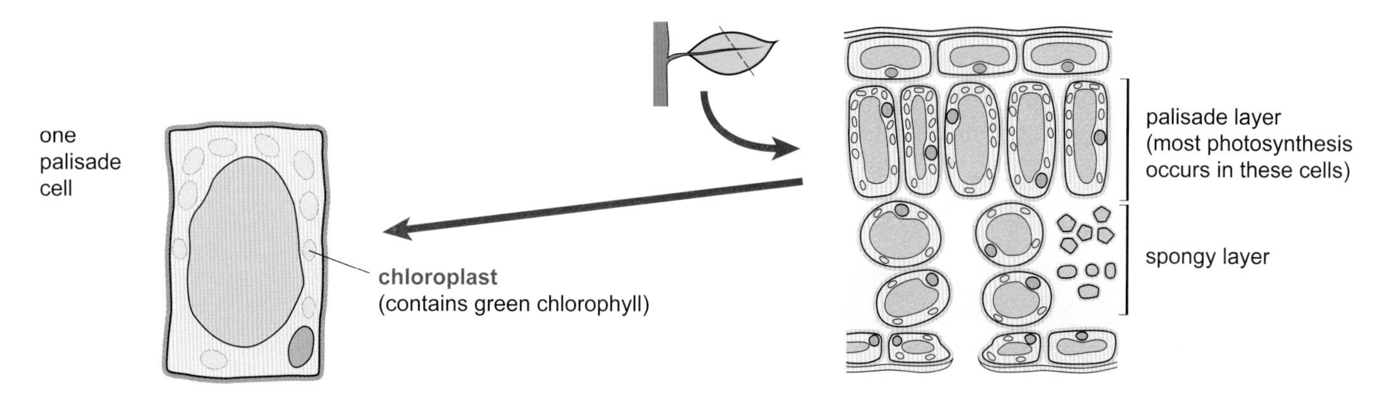

A slice across a leaf, showing the structure inside.

Glucose – the first product of photosynthesis

Plants use glucose for:

- making starch – they store the starch in roots, stems, leaves, fruits and seeds;
- releasing energy in respiration – they use this energy for their life processes;
- making new materials such as **proteins** and oils.

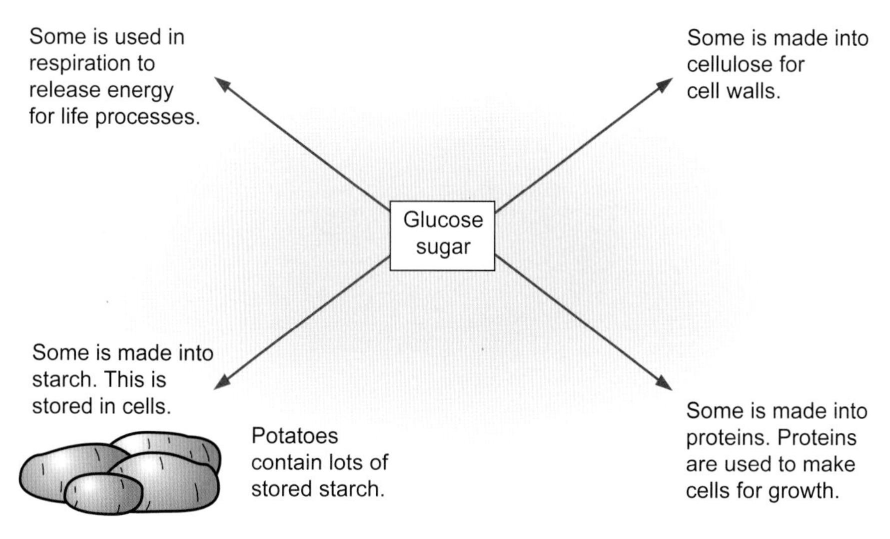

What happens to the glucose made during photosynthesis?

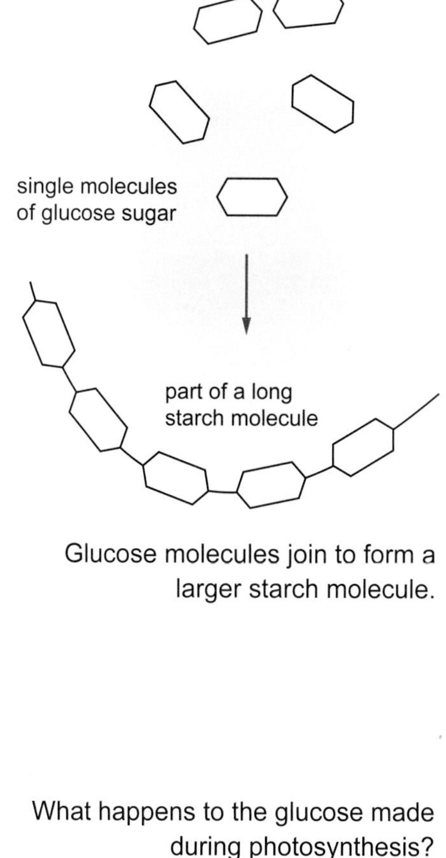

Glucose molecules join to form a larger starch molecule.

So, some of the glucose is used to release energy, while the rest increases the **biomass** of the plant.

Question 6

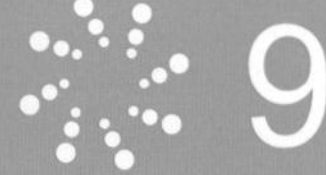

9C.3 Roots, water and minerals

You should already know

Outcomes

Keywords

Plant roots are important for anchorage and because they take up water from the soil. On the diagram of the plant, look for **root hairs** just behind the tips of roots.

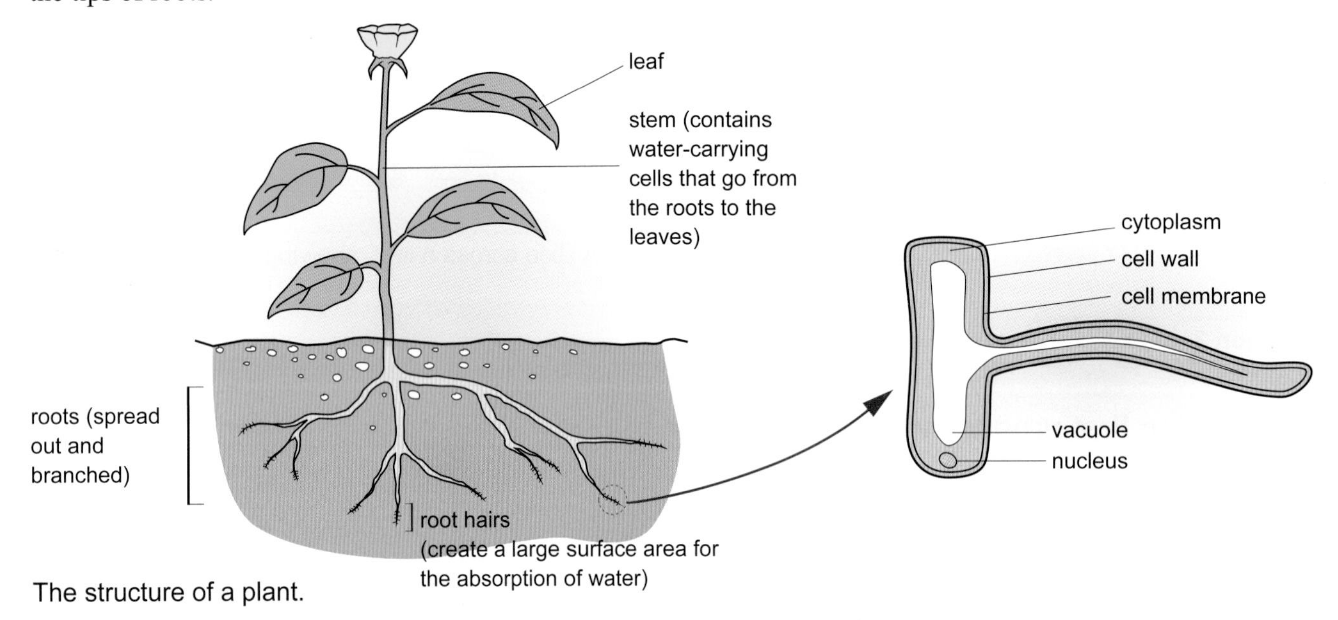

The structure of a plant.

Question 1 2 3 4

Root hair cells, like other living cells, need oxygen for **respiration**. The oxygen is needed to release energy from food for the cell's life processes.

Question 5

Many plants cannot survive in flooded or waterlogged soil. There is less oxygen dissolved in water than there is oxygen in air.

Antonis is a farmer in a small village called Livadia on the south coast of Cyprus. Livadia is in a valley between the coast and the hills. For many years, when Antonis was a young man, his crops were ruined when the valley flooded. Nowadays, man-made rivers take any flood water away from the valley and into the sea.

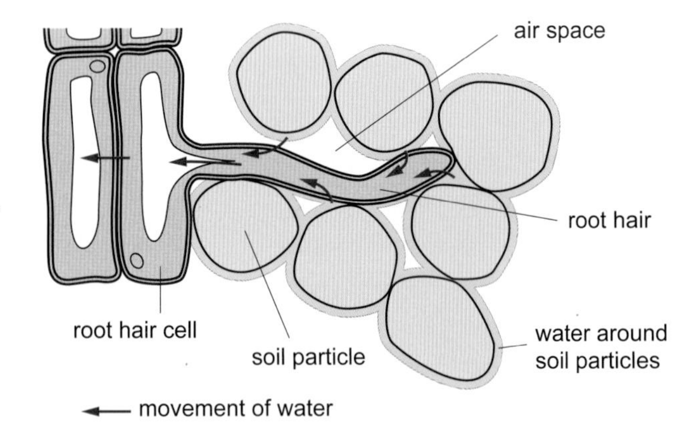

A root hair cell in the soil.

Question 6 7

How plants use water

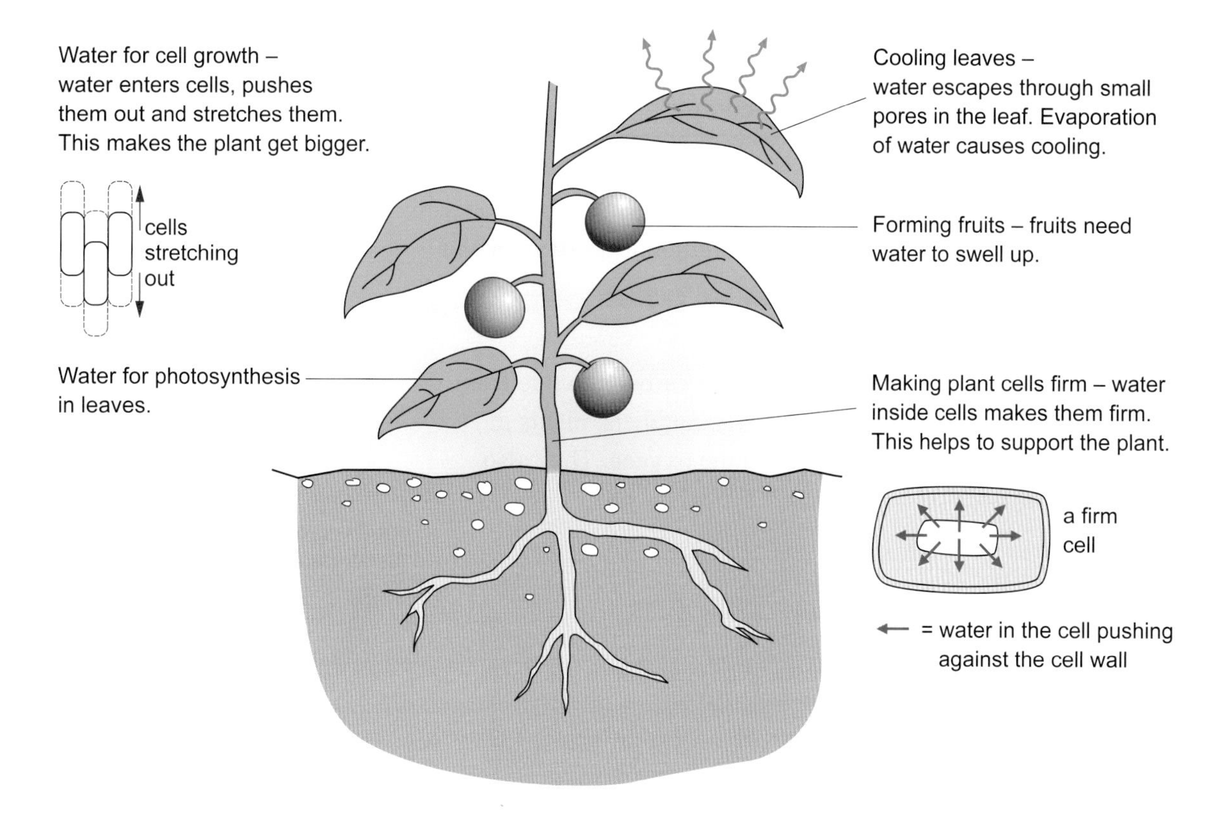

Question 8

The use of minerals in plants

The water in soil has **minerals**, such as **nitrates**, dissolved in it. Plants need nitrates to make proteins. They need proteins to make and repair cells.

Look at the pictures of plants growing in solutions of minerals.

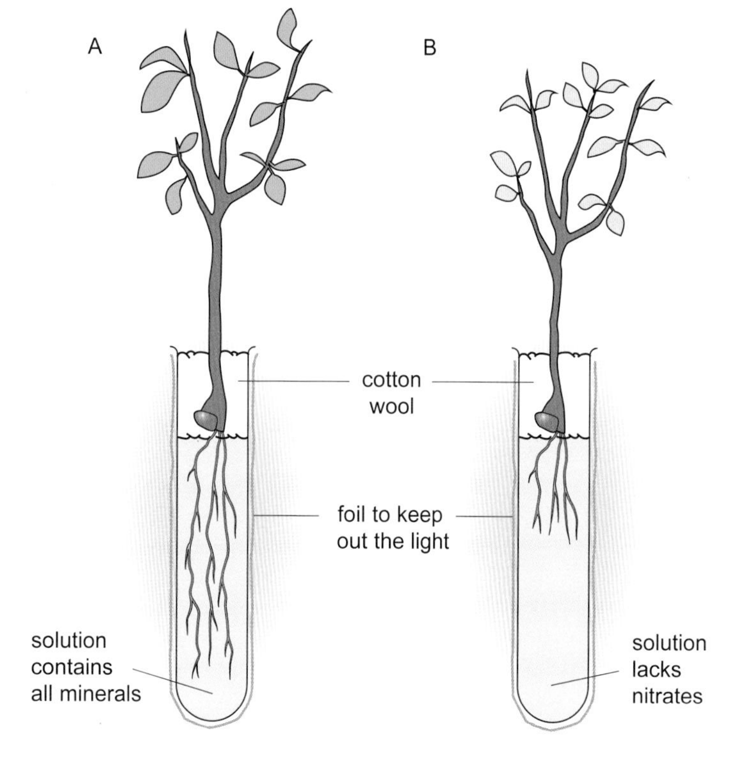

Question 9 **10** **11** **12**

9C.4 Green plants and the environment

You should already know

Outcomes

Keywords

Millions of years ago, there was no oxygen in the Earth's atmosphere. Then green plants began to add oxygen. Now, the amounts of oxygen and carbon dioxide remain more or less balanced. Oxygen makes up about 21% of the atmosphere and carbon dioxide about 0.03%.

When they photosynthesise, all green plants take in carbon dioxide from the air and give out oxygen. Because of their large numbers, plants in tropical rainforests and oceans produce most of the oxygen. They also remove carbon dioxide from the air.

Look at the equations.

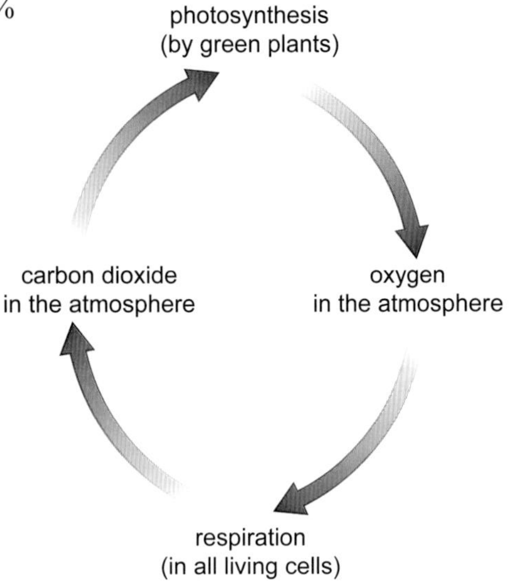

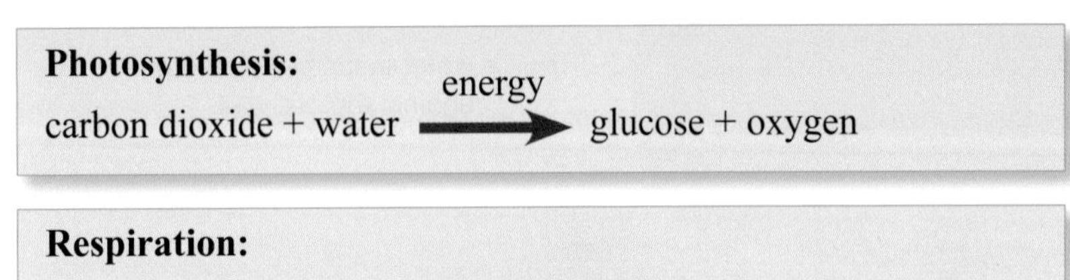

Respiration:

glucose + oxygen ⟶ carbon dioxide + water + energy

Question 1

Altering the balance

As we cut down more forests and pollute more of the oceans, less carbon dioxide is removed from the atmosphere by plants. Also, the burning and rotting of wood increases the amount of carbon dioxide in the atmosphere.

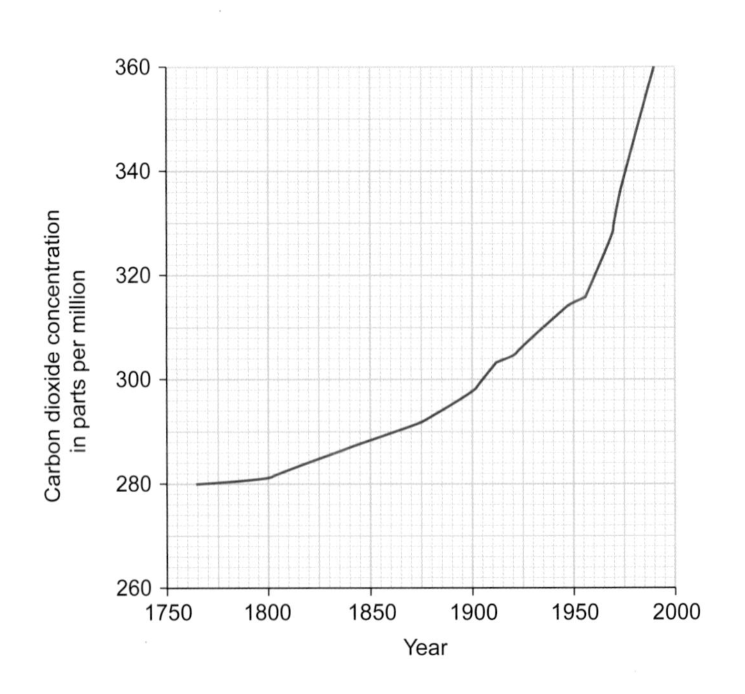

Question 2

Greenhouse gases

Carbon dioxide is one of a number of gases called **greenhouse gases**. Greenhouse gases stop heat escaping into space. Many scientists believe the average temperature of the Earth is now increasing. This is called **global warming**.

Methane is another greenhouse gas. It is produced in landfill, marshes and rice fields, and by animals such as cattle.

Question 3 

Over millions of years, the temperature of our planet has often changed. For example, dust in the atmosphere from volcanic eruptions has sometimes caused cooling.

Question 5 

Why we need to protect the environment

Conservation is the protection of the environment and of living things. We need to conserve forests, oceans and other places where plants and animals live.

For example, forests:

- help to maintain the balance of gases in the atmosphere.
  This protects us from climate change.
- provide us with timber, fruits, nuts, medicines and rubber.
  By managing forests carefully, they can keep on doing so.
  We say that they are **sustainable**.
- provide a home for many living things.
  When we cut down forests, plants and animals lose their homes.
  The variety of living things decreases.
  We say that **biodiversity** decreases.

We need to conserve our forests. Only about 10% of the UK is covered by trees.

Question 7

Review your work

Summary ➡

9C.HSW Food or fuel – a scientific argument?

You should already know

Outcomes

Keywords

You have seen in this unit, examples of how the scientific process of **investigation** provided answers to questions about:

- the chemistry of photosynthesis;
- the balance of gases in the atmosphere.

Remember

- **Photosynthesis**
  carbon dioxide + water $\xrightarrow{\text{energy}}$ glucose + oxygen
- **Respiration in all living things** (also the equation for burning)
  glucose + oxygen $\longrightarrow$ carbon dioxide + water + energy
- We call the mass of the materials in living things biomass.
- Plants are producers of biomass and help to maintain the balance of oxygen and carbon dioxide in the atmosphere.

We can use plant **biomass** for food and fuel. You have seen that fuels such as wood from plants are called biomass fuels. We can also use food crops to make fuels for engines. We call them **biofuels**.

Plants take carbon dioxide out of the air. So using biofuels raises the carbon dioxide concentration of the air less than using petrol or diesel. But studies have shown that some biofuels are responsible for more carbon emissions than others. It depends on the amount of energy used to grow, transport, harvest and process the crop.

Plant product	Food use	Biofuel use
sugar	cakes, sweets	ethanol (alcohol)
rapeseed, corn and palm oil	cooking oil, margarine	biodiesel

Question 1 2 3

If we use food products as fuel, there will be less food available.

Target of using 40% of US corn production for biofuel by 2017

Riots in Mexico as price of corn rises by 30% (2007)
Corn, including corn tortillas, is a basic food in Mexico …

Question 4

We can do the science, but how should we act?

We can use the scientific process to answer questions such as:

Which crops grown for biofuel reduce carbon emissions?

How can we increase the yields of crops?

Scientists have found ways of increasing **crop yields** using:

- fertilisers, pesticides and better cultivation methods;
- **selective breeding**, including finding strains suited to growing in particular places;
- **genetic engineering**.

But we can't use the scientific process to answer questions such as:

Should we use more food crops for fuel?

Should we replace more forests with crops?

It can only predict what might happen if we do. These are questions of **ethics**, economics and politics.

Question 5

The EU (European Union) policy for road transport is for 10% of vehicles to be run on biofuels. But as we use more crops for biofuels instead of food, problems are arising. These include:

- even greater shortages of food in the developing world;
- rising food prices worldwide;
- rainforest destruction. For example, in Indonesia:
  - they are cutting down rainforest to grow more palm oil;
  - plants and animals are losing their habitats;
  - **biodiversity**, the variety of plants and animals, is falling;
  - people are being forced off their land.

When we use land in the developing world for growing crops for fuel, the people there have even less land to grow food.

We must not create new environmental and social problems in our attempts to prevent climate change.

As humans use more land for crops, habitats are lost. So, many species of plants and animals die out.

Cutting down trees affects the environment. Rainfall may be reduced and soil can be damaged and eroded.

Question 6

So the EU is reviewing this policy. Environment Commissioner Stavros Dimas said that it would be better to miss the target than to reach it by harming the poor or damaging the environment.

Question 7

9C.1

1. The word 'biomass' is made up of the words 'bio' and 'mass'.
   Use a dictionary to find out what these two words mean.
2. **a** Suggest why van Helmont thought that water is enough for plants to grow.

   b What is the evidence that a plant's food does not come from the soil?
3. Write a few sentences to explain to Eleanor how Mr Holmes' experiment provides evidence that a plant's food does not come from the soil.
4. **a** Describe the changes in the amount of carbon dioxide over 24 hours.

   b Suggest an explanation for these changes.
5. Use the information in the diagram to write a word equation.
6. Suggest a control for the experiment with *Elodea*.
7. Look at the graph.
   Explain what it shows.
8. Suggest what happens when this experiment is done in the dark.
   Explain your answer.
9. Describe an example from this Topic of a scientific investigation producing new evidence leading to a new explanation.

9C.2

1. Look at the diagram.
   What can you conclude from this experiment?
2. What evidence is there from the experiment that chlorophyll is needed to make starch?
3. Describe the distribution of chloroplasts in the leaf.
4. Describe the shape of the palisade cells.
5. Suggest why most of the chloroplasts are nearer the top of the palisade cells.
6. Use the information in the diagram to complete a table.
   Use these headings for your table:
   - Substance made from glucose
   - How the plant uses this substance

9C.3

1 Describe the route that water takes in a plant from the soil to the leaves.

2 What do the leaves use the water for?

3 Look at the diagram.
Write down two things that give roots a larger surface area for absorbing water.

4 How does the shape of the root hair cells help the plant take up water?

5 Where do root cells get their oxygen for respiration?

6 Why did the flooding kill Antonis' crops?

7 Write down two things that roots take from the soil.

8 Use the information in the diagram to draw a table.
Use these headings in your table:
- Use of water
- How does it happen?

9 The experiment was set up to find out the effects of a lack of nitrates.
Why was bottle A needed?

10 Explain why you need to set up ten plants like this in each solution, not just one of each, as shown.

11 Describe and explain the effects of a lack of nitrates on the plant.

12 Find out how farmers get extra minerals into the soil for their crops.

9C.4

1 Make a list of differences between the equations.

2 How much did the amount of carbon dioxide in the atmosphere increase between 1960 and 1990?

3 Write down the name of <u>one</u> greenhouse gas other than carbon dioxide.

4 Greenhouse gases are produced by or for each one of us.
In a small group, discuss what <u>you</u> can do to reduce the amount produced directly or indirectly by you.

5 Some people think that human activities are the cause of global warming. Others disagree.
Suggest why this is a difficult question to answer.

6 What types of evidence are scientists using to work out a link between global warming and what humans are doing?

7 Find out:

a the name of a place in the world where people are cutting down forests;

b why they are cutting the trees down there;

c what the effects on their environment are.

9C.HSW

1 Write down the names of two biofuels.

2 a Some people say that burning biomass fuels does not add extra carbon dioxide to the air.
Use the equations to help you to explain this.

b Others say that the fuel used to grow, harvest, transport and process a crop means that we are cutting carbon dioxide emissions by less than we think.
In your group, make a list of the energy inputs into producing biofuels.

3 The Royal Society is a famous group of scientists. Its report on biofuels in 2008 suggested that each biofuel should be assessed and that we should use only those that really do save carbon emissions.
In your group, agree on a way of finding out whether or not a particular biofuel crop from a particular place saves carbon emissions.

4 Look at the newspaper headlines.
Suggest why the US policy on biofuels affected the price of corn in Mexico.

5 The scientific process cannot answer the question of whether or not we should replace forests with biofuel crops.
In your group, discuss:

a Why the scientific process can't answer that question.

b Who does have responsibility for answering that question.

c Ways you can influence the decision-making process.

6 Look at the speech bubbles.
Explain the sentence from the left-hand speech bubble to a ten-year-old.
Include examples of social and environmental problems in your explanation.
(Hint: use examples from the text and other speech bubbles.)

7 Some people think that increasing the use of biofuels will do more harm than good.

a Work in a group to suggest arguments for and against this opinion.

b For each argument, agree whether it is a scientific, an economic, an environmental or an ethical argument, or more than one of these.

Scientific enquiry

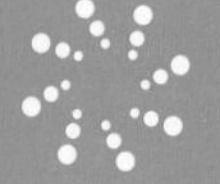

9D.1 Where does our food come from?

You should already know | Outcomes | Keywords

Your food supplies you with:

- energy for all your life processes;
- materials for growth and repair.

Only green plants make food. That is why we call them **producers**. Animals eat or consume food. So they are called **consumers**.

Remember

carnivore	secondary consumer
↑	↑
herbivore	primary consumer
↑	↑
green plant	producer

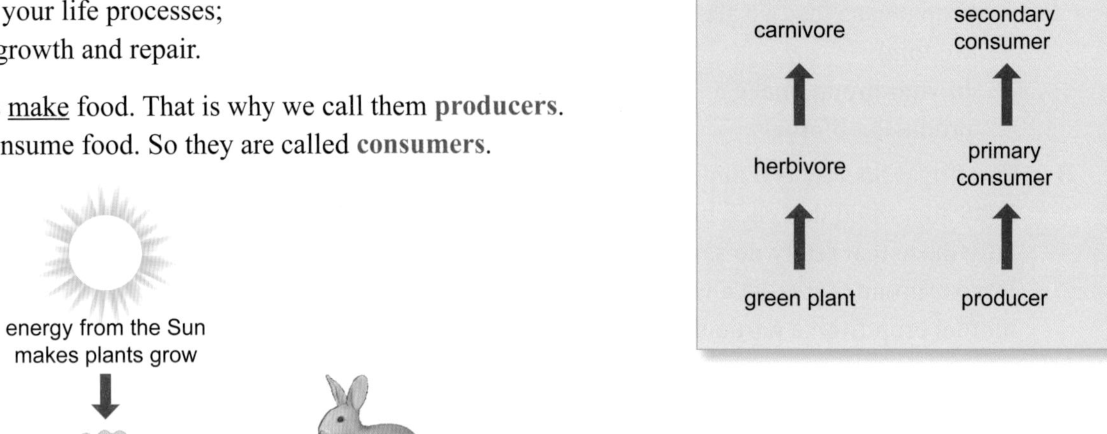

Plants use sunlight energy to make food in **photosynthesis**. So, the energy stored in food comes from the Sun.

Question 1 2 3

Plants make glucose first, and then they change some of it into other foods.

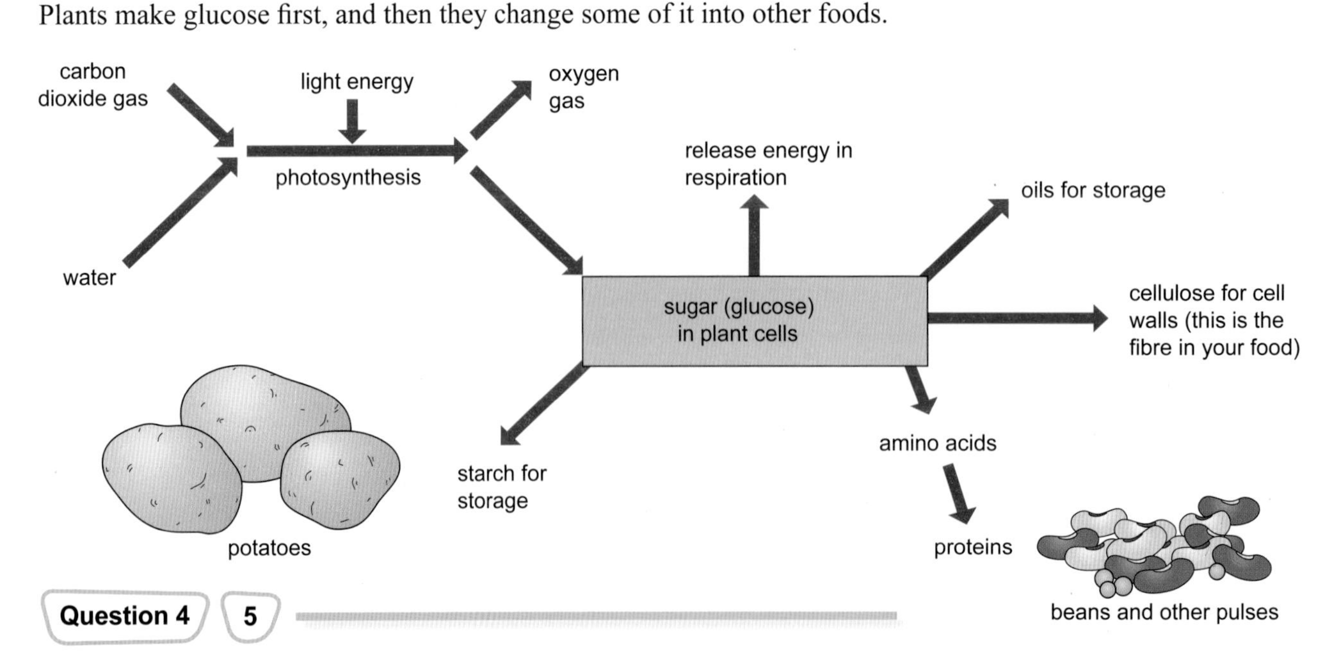

Question 4 5

A lot of our food is processed. Sometimes it is hard to tell which animals and plants it came from. This food web shows where the ingredients in a chicken pie come from.

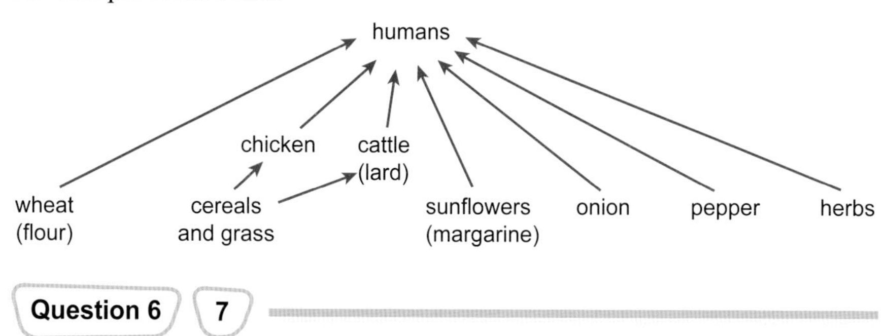

Ingredients

chicken	lard	pepper
onion	margarine	salt
flour	herbs	

Question 6 | 7

Most of us don't grow or catch and kill our own food. Farmers in the UK and around the world grow the food we eat.

We eat different parts of different plants

Sometimes we eat one part of a plant but not another. Some parts are nicer to eat than others. You can eat an apple, but not the bark of the tree.
In the food web for the chicken pie above:

- flour, pepper and cooking oil come from seeds;
- onions are bulbs;
- herbs are usually leaves.

Some parts of plants contain more **nutrients** than others. Like us, plants release energy from carbohydrates and fats in respiration. They store food so that they can survive the winter and grow new leaves in the spring. So we eat the food that plants make and store for their own benefit, not ours.

Question 8 | 9  | 10 | 11

	Root	Stem	Leaves	Leaf stalk	Flower	Fruit or seed
beans						•
broccoli					•	
cabbage			•			
carrot	•					
celery				•		
rice						•
sugar cane		•				
tomato						•
wheat						•

Fruits and seeds contain proteins for growth and carbohydrate or fats for energy. When humans and other animals carry away and eat the soft, sugary parts of fruits, they are helping to spread the seeds that are inside the fruits.

Question 12

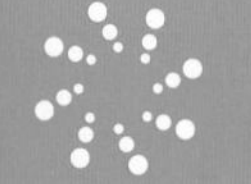

9D.2 How do fertilisers affect plant growth? (HSW)

You should already know | **Outcomes** | **Keywords**

The equation shows the raw materials and the products of photosynthesis.

carbon dioxide + water —(energy from sunlight)→ sugars and oxygen

To make proteins, plants use sugars and minerals such as **nitrates** too. **Minerals** are plant **nutrients** that plants take into their roots from the soil.

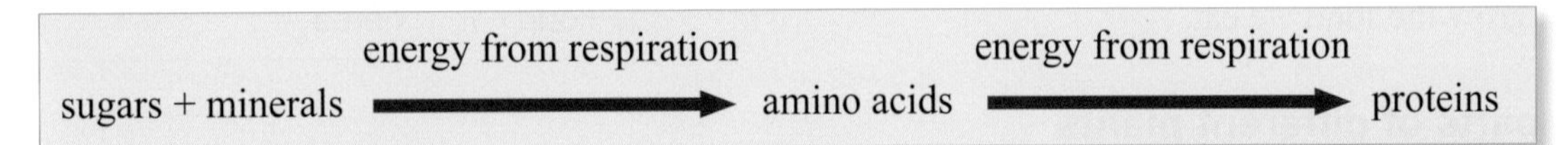

sugars + minerals —(energy from respiration)→ amino acids —(energy from respiration)→ proteins

Plants also need small amounts of minerals to make **chlorophyll** and other chemicals.

Question 1  2

When farmers and gardeners harvest plants, they take away the minerals stored in them.

So to make sure that the next crop gets enough minerals, they add **fertilisers** to the soil.

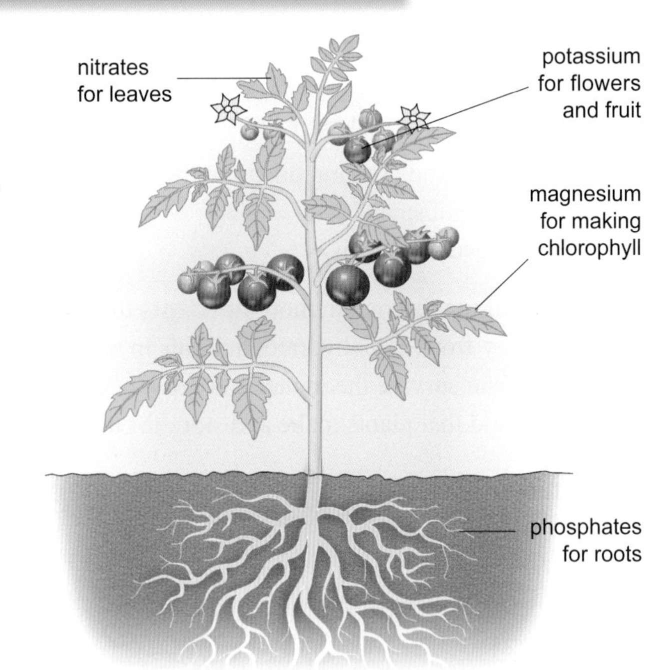

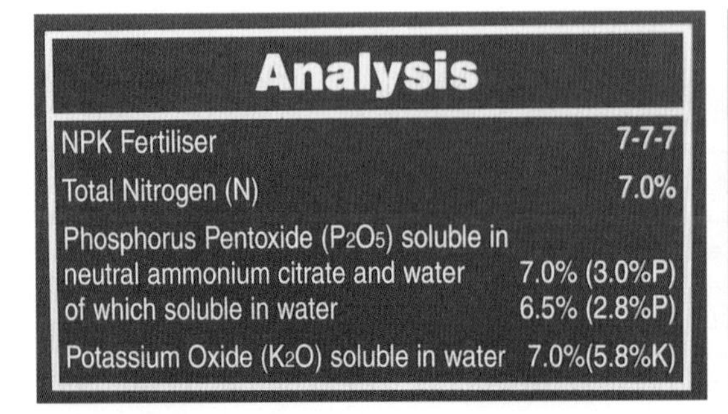

Analysis

NPK Fertiliser	7-7-7
Total Nitrogen (N)	7.0%
Phosphorus Pentoxide (P_2O_5) soluble in neutral ammonium citrate and water	7.0% (3.0%P)
of which soluble in water	6.5% (2.8%P)
Potassium Oxide (K_2O) soluble in water	7.0%(5.8%K)

The fertiliser contains the minerals that plants use most.

UK Fertiliser Declaration

EC Fertiliser
NPK Fertiliser 15-30-15 with Trace Elements.

Nitrogen (N) total	15%	Boron (B) soluble in water	0.02%
Ammoniacal nitrogen (N)	6.5%	Copper (Cu) soluble in water	0.07%
Ureic nitrogen (N)	8.5%	Iron (Fe) chelated by EDTA	
Phosphorus Pentoxide (P_2O_5)		soluble in water	0.15%
soluble in Neutral Ammonium		0.15% chelated by EDTA	
Citrate and in water	30% (13.1% P)	Manganese (Mn) chelated by EDTA	
of which soluble in water	30% (13.1% P)	soluble in water	0.05%
Potassium Oxide (K_2O)		0.05% chelated by EDTA	
soluble in water	15% (12.4%K)	Zinc (Zn) soluble in water	0.06%

This fertiliser contains extra minerals.

Question 3  4

Lack of a mineral

We can investigate the lack of a particular mineral by putting small plants in water or clean sand that contains no minerals. Then we can add different minerals to find out what effects they have on plant growth.

Helen and Vijay grew 20 cuttings in sand. They gave 10 plants a fertiliser containing all the minerals that plants need. For a second group of 10 plants, they left out just one mineral.

Plants A and B in the pictures are average plants from the two groups.

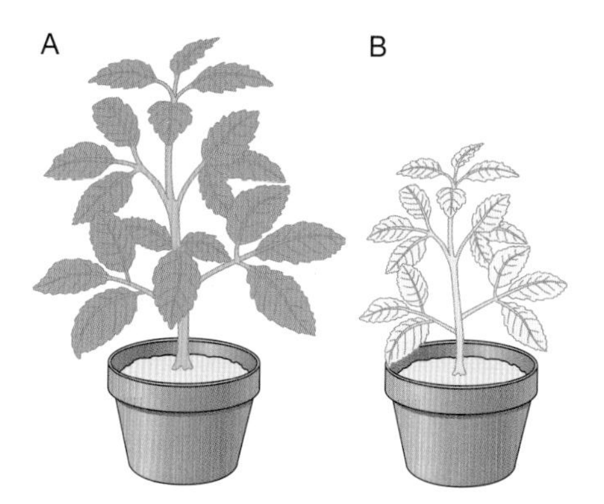

Plant A had all the minerals it needed.

Plant B lacked magnesium.

Question 5  6  7  8

Concentration of a mineral

We can find out how different concentrations of a mineral affect the **yield** of a crop. The yield is the amount of a crop that we can harvest.

Some students at an agricultural college used different amounts of nitrate fertiliser on wheat in different fields.

They mixed different amounts of fertiliser with the same amount of water for each hectare.

The graph shows their results.

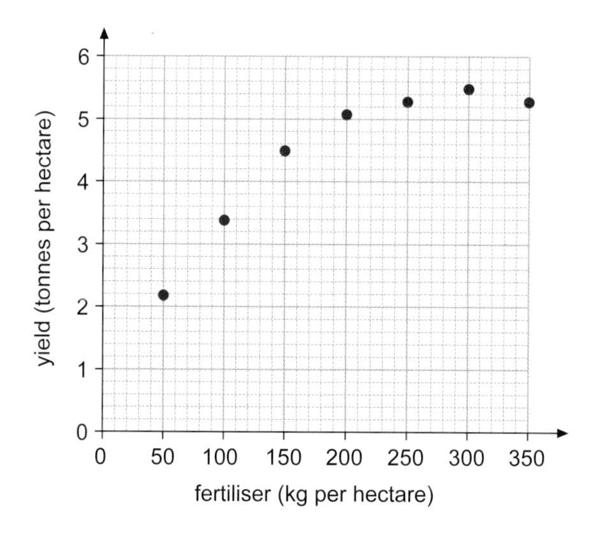

Spreading fertiliser on a wheat crop.

Question 9 10  11

When you investigate fertilisers yourself, remember some of the ideas in this topic.

Check your progress

9D.3 Plants out of place

You should already know | Outcomes | Keywords

What is a weed?

A **weed** is a fast-growing plant that is growing where it is not wanted.

Couch grass in a flower bed.

Daisies in a lawn.

Question 1 | 2

How do weeds affect crop production?

Weeds compete with crop plants. So they reduce the **yield** of crops.

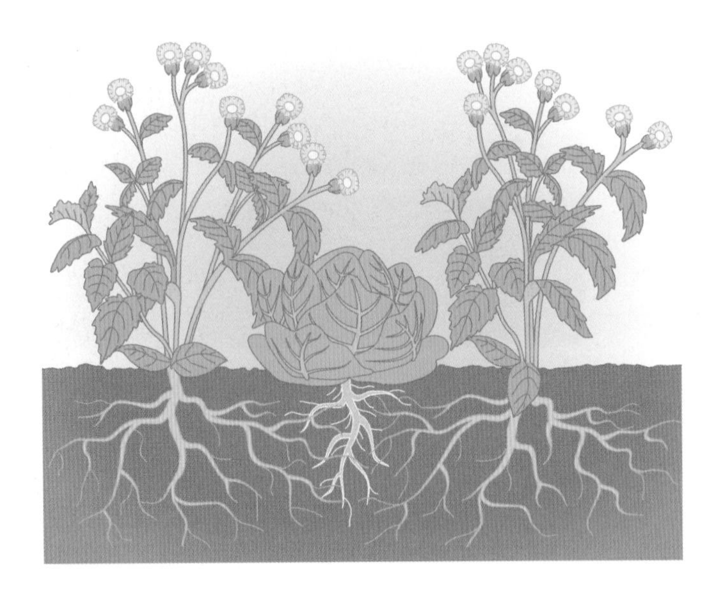

Weeds **compete** with the cabbage for light, water, minerals and space.

Question 3 | 4

How do we control weeds?

We can remove weeds by hand if crops are grown in rows that are far enough apart. This takes a lot of time and effort.

We can use weedkillers.

A **weedkiller** is a chemical that kills weeds. You have to choose a weedkiller that doesn't kill your crop plants as well as the weeds.

Look at the table.

Farm workers weeding a field of strawberries.

Weedkiller	What happens	Types of plants that it kills
glyphosate	plants wilt, leaves turn brown, then the plant dies	all plants – leaves and roots
2,4-D	leaf stalks and leaves grow unevenly, so they twist as they grow and the plant dies	broadleaved plants such as dandelion, plantain, daisy, clover and many crop plants, but not the grass family
diquat	leaves turn yellow, then black	most plants, but not those with deep roots

Question 5 6

Farmers use selective weedkillers to kill broadleaved weeds in their wheat fields. Look at the pictures of the wheat fields. A farmer treated the field on the right with weedkiller, but not the one on the left.

Question 7 8 9 10 11 

9D.4 Pests

You should already know | Outcomes | Keywords

These animals are **pests** because they eat farmers' crops, but they are also important food for other animals.

Mrs Gilbert has a field of cabbages. Last year, her crop was badly damaged by slugs. This year, she has decided to use a **pesticide** to kill the slugs. She has chosen to use slug pellets – a pesticide that it is claimed kills only slugs and snails and is safe for other wildlife.

Some common pests of crop plants.

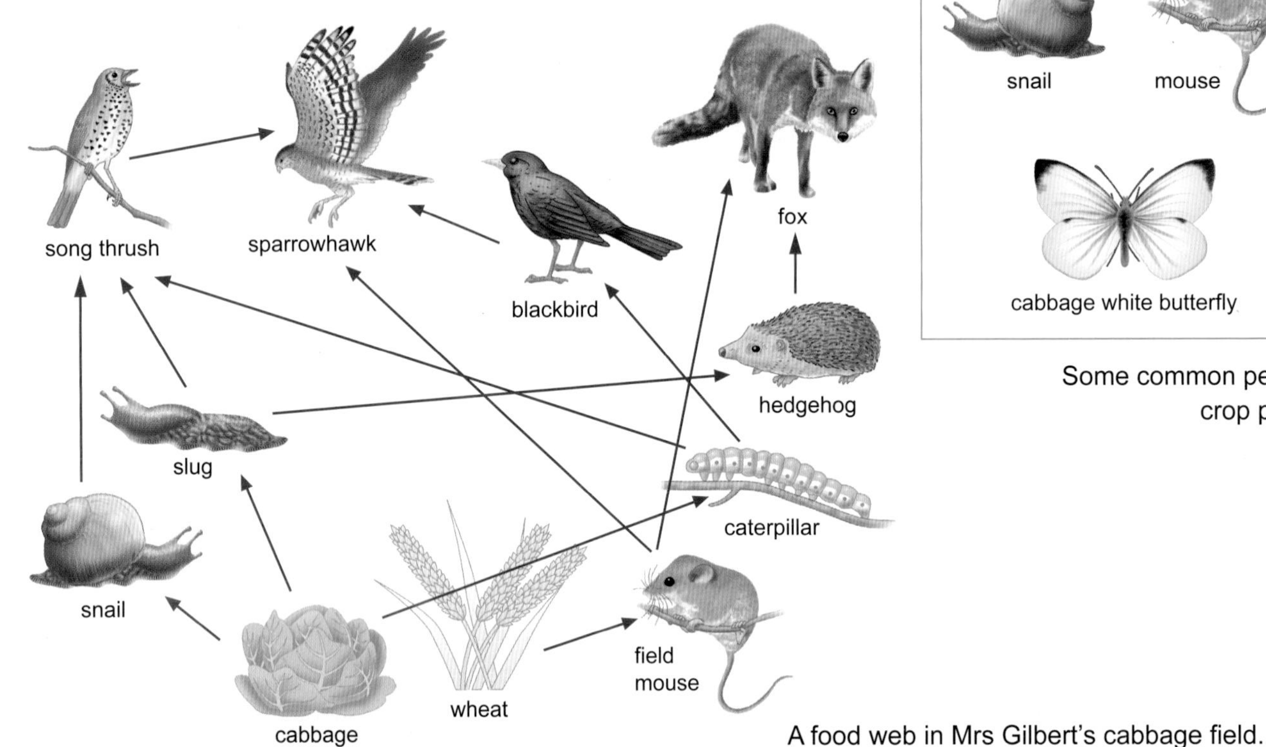

A food web in Mrs Gilbert's cabbage field.

Question 1 2 3 4 5

What are pesticides?

Pests can damage up to 40% of a crop. Farmers use pesticides so that they can produce:

- more food and make more profit;
- undamaged food, which will sell for a good price.

Pesticides kill other wildlife as well as pests.

Type of pesticide	What it kills
herbicide	plants
insecticide	insects
molluscicide	molluscs (slugs and snails)

Question 6 7

Not all insects are pests of crop plants. Useful insects, such as bees, pollinate flowers.

Question 8

Persistent pesticides

Some pesticides aren't broken down easily in the body or in the environment. We call these persistent pesticides. DDT is an example of a persistent pesticide that is now banned in Europe and many other parts of the world.

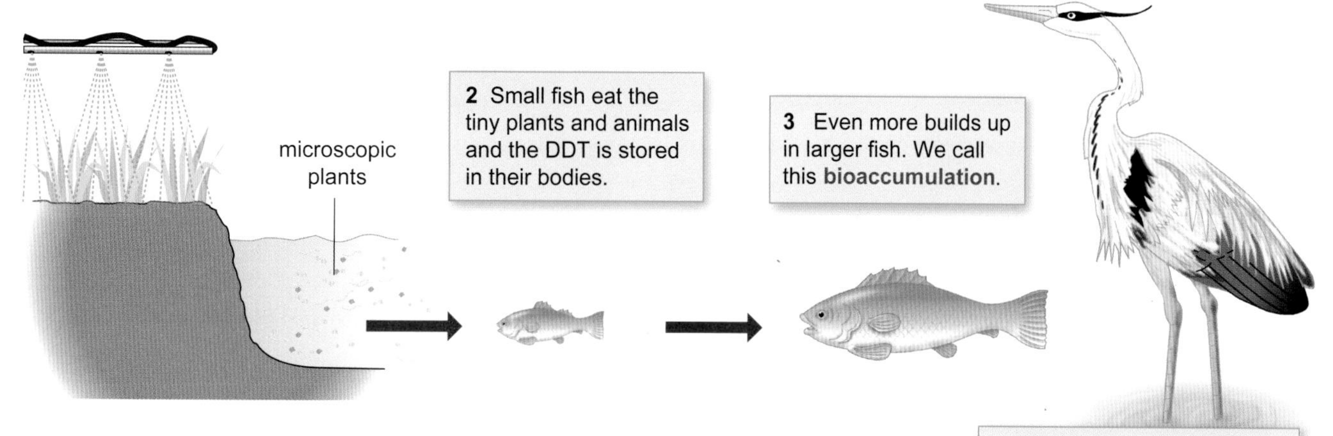

Question 9 10

Non-persistent pesticides

Pyrethroids are non-persistent pesticides; they break down into simpler substances in living cells and in the environment. We say that they are **biodegradable**.

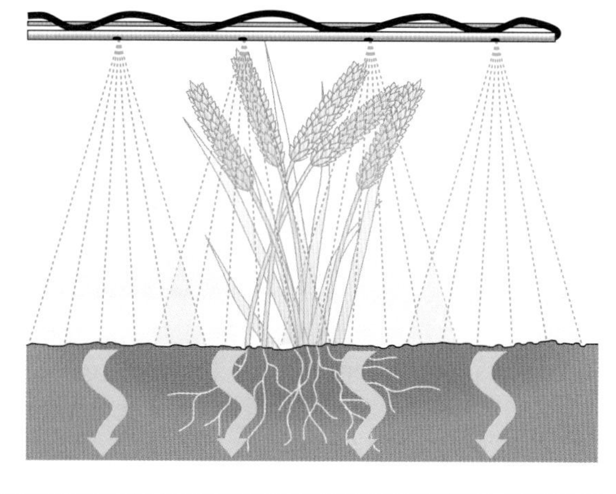

Farmers spray pyrethroids onto fields at a safe dose for wildlife. In the soil, they quickly break down into harmless chemicals.

Question 11

9D.5 Producing more food

You should already know | Outcomes | Keywords

How the environment affects plant growth

The diagrams summarise what you should know about the effects of the environment on plants.

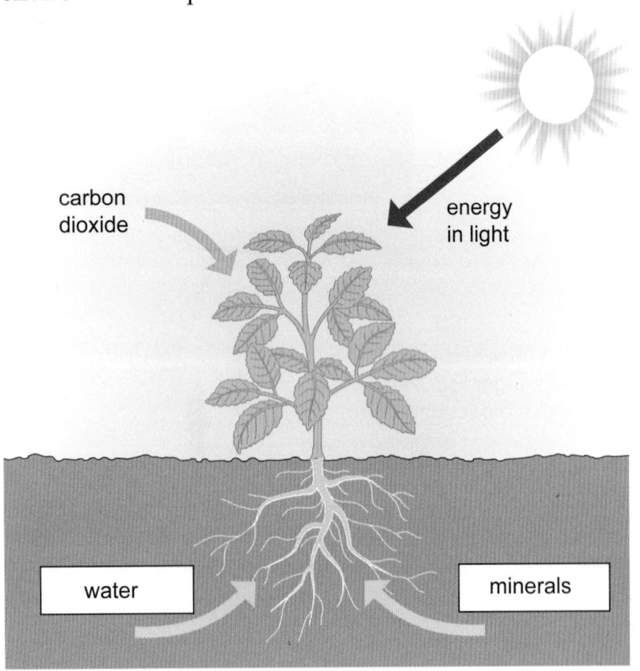

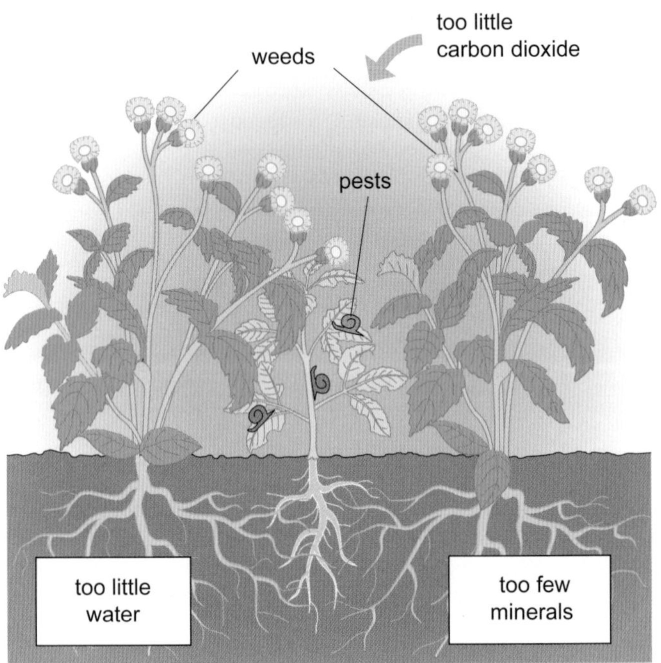

Question 1 | 2

Improving the environment for plants

Farmers and gardeners control some **environmental conditions** for their plants. They can buy **fertilisers** or grow plants in heated greenhouses, but it is too expensive to control everything.

Question 3

Better conditions mean better crops

The better the conditions for plants, the higher the **yield** of a crop. Environments vary. Some environments provide better conditions for plant growth than others.

Farmers try to improve the conditions. So farmers in different places need to change different things.

In Malaysia, the climate is wet, hot and sunny. Farmers use manure as a fertiliser and need to control pests, weeds and diseases.

Question 4 | 5 | 6 | 7

Can we produce more food?

As well as controlling the environment, farmers and scientists use **selective breeding** and other ways of producing plants that give better crops. (You have come across this idea before.)

But scientists think that the population of Earth will double by the year 2050. So even more food will be needed by then.

Egypt is dry, hot and sunny. Salim uses manure to provide plant nutrients, takes out weeds and sometimes uses pesticides. He pumps water into channels in his fields daily. We call this **irrigation**.

Question 8

One problem is that in other places good land is being destroyed by overuse. What we need are ways of continuing to produce more food without damaging the Earth. This will let development continue, so we call it **sustainable development**.

Question 9

In commercial greenhouses in the UK, growers control pests and diseases and the amounts of water and fertilisers. Opening and closing of windows is automatic.

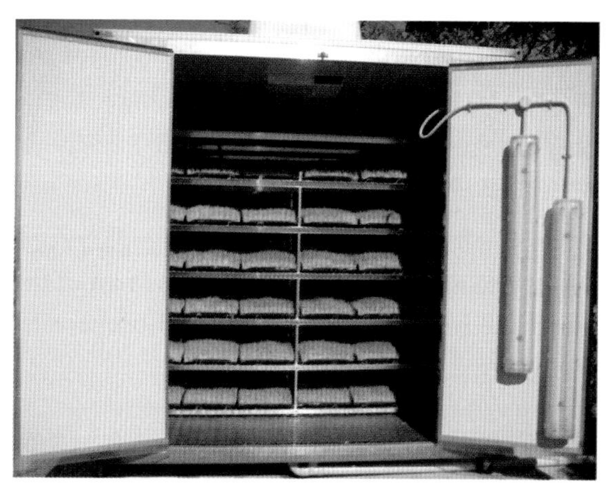

Some farmers use fodder units to grow barley 'grass' from seed. It takes 2 or 3 litres of water to produce each kilogram of fodder rather than the 80 litres it usually takes.

Review your work

Summary ➡

You should already know

Outcomes

Keywords

Scientists investigate the environment using various kinds of **models**.

Scientists can use models to help them to answer some questions.

Models on paper

You have used food webs and biomass pyramids to model feeding relationships in a **habitat**. You can model processes in an **ecosystem** using cycles such as the carbon cycle. But a whole ecosystem includes all the interactions between the environment, plants, animals and micro-organisms. The diagram is a model on paper of some of the main interactions.

Question 1 2

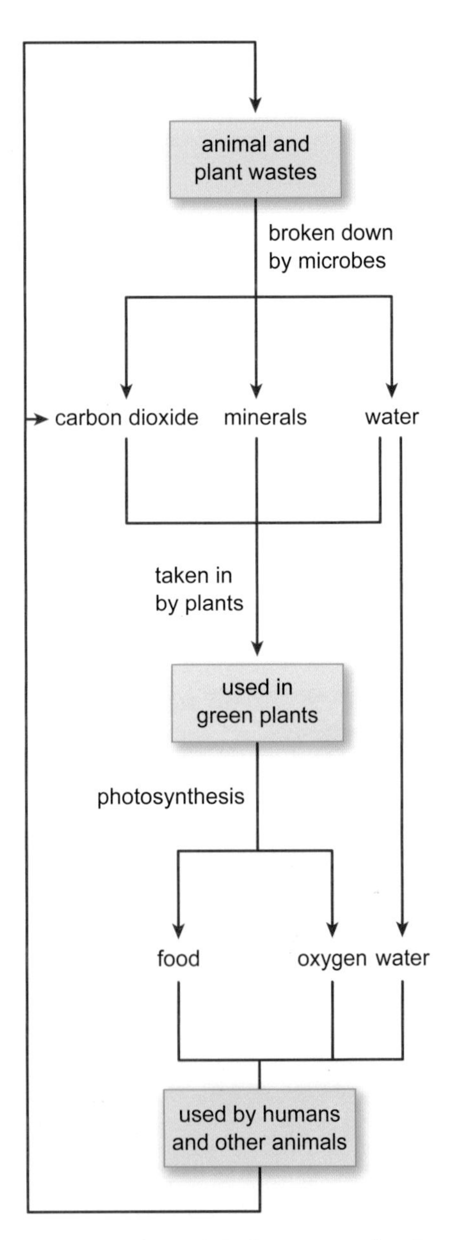

A model of some cycles in an ecosystem.

Ecosystem models

If you have ever tried to keep a small-scale model such as an aquarium going, you will know how hard it is to get the right balance. So it is not surprising that scientists in Arizona, USA, had a problem when they tried to model a whole space station. They used their knowledge to predict what was needed.

As their model, the scientists built a glass and steel building. Unless a planet has an Earth-like atmosphere, a space station has to be a sealed system. So they could allow in only light from a 'sun'. They put in systems to **recycle** all carbon dioxide and other waste. So their model is also a model for the Earth.

To test their predictions, they sealed eight people and about 4000 different plants and animals inside. We call the part of the Earth inhabited by living things the **biosphere**. The Earth is Biosphere 1, so they called their model Biosphere 2.

Question 3   5 6

Continuing research in Biosphere 2

Scientists planned to use Biosphere 2 for 100 years. Apart from leaks, everything has to go in and out through an airlock. So Biosphere 2 is an excellent model for investigating changes in a closed system.

The first group of people stayed in Biosphere 2 for two years. Unfortunately, they had less food and oxygen than normal. They lost weight but their health improved in many ways.

As there were problems, scientists had to make some changes to their model.

Biosphere 2.

Question 7

Surprisingly the atmosphere leaked about 30 times less than the Space Shuttle – but still by 11% per year.

When the first crew left, scientists spent 5 months:

- repairing leaks and sealing concrete so that it could no longer absorb carbon dioxide;
- collecting data such as changes in populations;
- updating systems for computers, videos, keeping air healthy, environmental monitoring, etc.;
- putting in extra lighting to improve crop growth;
- planting extra food plants in the rainforest
- introducing different living things, including:
  - **predators** of pests, e.g. geckos and toads to eat the cockroaches;
  - improved varieties of crop plants;
- setting up new research projects.

The next crew for Biosphere 2 was of six people instead of eight. But there were plans for other scientists and medical doctors to spend time there.

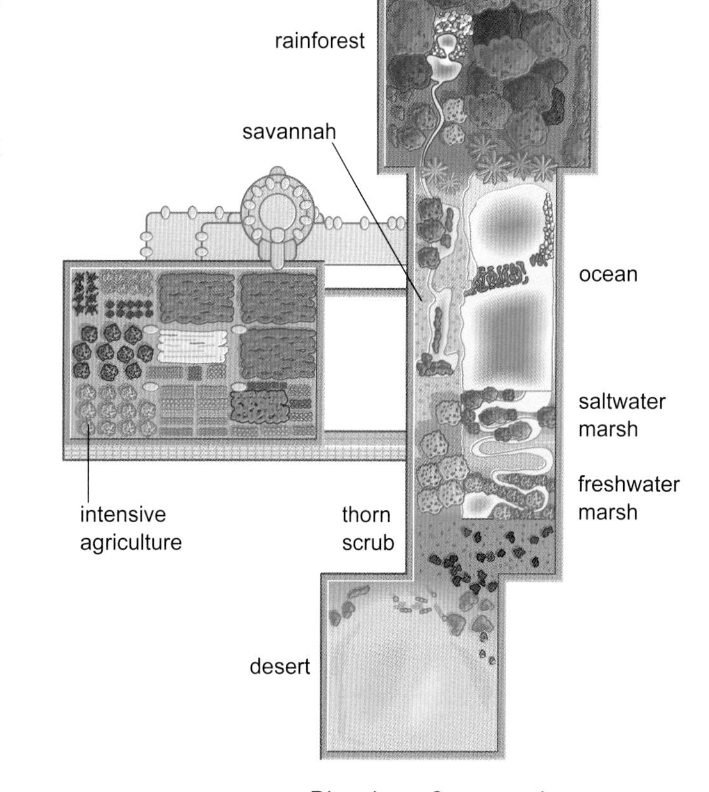

Biosphere 2 covers three acres – a little more than two football pitches.

Question 8 

Computer models

Some scientists use computer models to predict what will happen in the environment. For example, there are computer models for predicting the weather and the effects of global warming. Scientists are continually improving these models, but they are not yet reliable. There are too many variables and unknown factors for the predictions to be reliable.

Question 13

9D.1

1 Write down <u>two</u> examples of food from:

   a green plants;

   b animals.

2 If it were not for the Sun and green plants, you'd have no food. Explain why.

3 Are you feeding as a primary consumer or a secondary consumer when you feed on:

   a green plants?

   b herbivores?

4 Plant cells use sugars in respiration. Why do plant cells respire?

5 Why do plants make proteins?

6 Look at the ingredients in the chicken pie. Which ingredients come from:

   a green plants?

   b animals?

7 Draw a food web, like the one for the chicken pie, for your favourite meal.

8 a Name <u>one</u> food plant.

   b Write down <u>one</u> part of that plant that we don't eat.

9 Write down <u>two</u> nutrients that we get from seeds.

10 Why do seeds contain proteins and an energy source?

11 The fleshy part of fruit, such as plums, contains sugars. The plum tree doesn't use these sugars in respiration. Animals eat and use them. Explain how the tree benefits from the sugars in plums.

12 The table shows some parts of plants that we eat. Draw a similar table for the following foods.

apple coconut grape lettuce maize cauliflower

rhubarb onion potato parsnip soy bean

9D.2

1 Write down <u>two</u> reasons why plants need minerals.

2 Write down the minerals that plants need for healthy growth of:

   a leaves;

   b roots.

continued

3 a Write down three minerals that both fertilisers supply.

b Write down three extra minerals in the second fertiliser.

4 Write down two things that you think farmers and gardeners take into account when they choose a fertiliser.

5 Why did Helen and Vijay grow their plants in sand?

6 Explain why they used 20 plants, rather than just two.

7 Suggest two other ways to make Helen and Vijay's test fair.

8 What do you think plants use magnesium for?
Explain your answer.

9 Why do you think the students dissolved the different amounts of fertiliser in the same amount of water?

10 Describe what the students found out.

11 The highest yield of wheat was at 300 kg of fertiliser per hectare. But the students decided that using 200 kg per hectare was more sensible.
Suggest a reason for this.

9D.3

1 In which photograph is the grass the weed?

2 In which photograph are the flowers the weeds?

3 What do the roots of weeds and crops compete for?

4 What do the leaves of weeds and crops compete for?

5 Couch grass has underground stems and deep roots that are very hard to kill.
Which type of weedkiller would you use to get rid of couch grass?
Explain your answer.

6 From the table, which weedkiller would you choose for getting rid of daisies in a lawn?
Explain your answer.

7 Why is it difficult to weed a field of wheat by hand?

8 Look at the pictures.
Suggest which field farmers prefer.
Explain your answer.

9 Conservationists are people who want to preserve wild plants and animals.
Suggest which field conservationists prefer.
Explain your answer.

10 Birds feed on the seeds of some weeds.
What can happen to the population of birds when farmers use weedkiller?

11 In a field of crops, nettles are weeds. But they are food for several animals.
Find out the name of an animal that eats nettles.

9D.4

1 Humans usually treat caterpillars as pests.
Name two animals which find caterpillars useful.

2 Look at the food web and describe how you think using slug pellets will affect the population of:

- **a** snails;
- **b** song thrushes;
- **c** sparrowhawks.

3 If slugs and snails are killed, the cabbage white caterpillar population will change.
Explain why the caterpillar population might go:

- **a** up;
- **b** down.

4 How can you explain to Mrs Gilbert that, even if the claim that slug pellets kill only snails and slugs is true, other wildlife will still be affected?

5 Find out if slug pellets really are safe for other wildlife.

6 What do you think '-cide' means at the end of a word?

7 Make up a name for a pesticide that kills rodents such as rats and mice.

8 Farmers don't spray their crops with insecticide when the crops are flowering.
Why do you think this is?

9 Look at the diagram.
How did DDT kill herons when it was sprayed on crops in dilute 'safe' amounts?

10 Why do you think it was 10 years before scientists realised there was a problem with DDT?

11 Why do farmers now use non-persistent pesticides?
Explain as fully as you can.

9D.5

1 Write down five environmental conditions that affect plant growth.

2 What is the source of energy for plant growth?

3 Look at the diagrams again.
Write a list of substances that farmers might need to add to make their crops grow faster.

4 Why are outdoor crops larger and faster growing in Malaysia than in:

- **a** Egypt?
- **b** the UK?

5 Tomato growers in the UK heat their greenhouses.
Write down one benefit and one disadvantage of doing this.

6 In the UK, crops such as wheat and beans grow faster in higher temperatures.
Why don't we provide higher temperatures for them?

7 Find out some benefits and some disadvantages of growing crops in greenhouses.

continued

8 Compared with growing fodder barley in open fields, explain why fodder units:

a produce fodder more quickly;

b use much less water.

9 Write down two reasons why we need scientists to carry on researching crops and crop production.

9D.HSW

1 Food chains and webs model feeding relationships.
Explain why food webs are the better models.

2 The diagram is a model for the recycling of materials. It is a better model than the carbon cycle for what happens in an ecosystem such as a pond or a space station.
Explain why.

3 Most aquaria are unbalanced systems. So you have to do things to keep them going.
In your group, think of what you have to do.

4 The Biosphere 2 building is a bit like a greenhouse. It lets in lots of light.
Explain why having a lot of light is important.

5 Work in a small group using the diagram as a model. Agree on how people on a space station might meet their needs for:

a food;

b water;

c oxygen;

d getting rid of waste;

e getting energy for cooking, etc.

6 In your group, make a list of the specialists that would have worked together to plan Biosphere 2.

7 In your group, suggest why there were bound to be problems with Biosphere 2.

8 What changes did the scientists make to increase food supplies?

9 Concrete absorbs carbon dioxide.
Suggest why the scientists wanted to prevent this.

10 Suggest where scientists put the extra lighting, and why.

11 In your group, discuss the advantages and disadvantages of using biological control of pests, such as using geckos to eat cockroaches instead of chemical pesticides.

12 Suggest why the second crew of Biosphere 2 was six people instead of eight.

13 It is hard to predict the weather, even using computer models.
Suggest why.

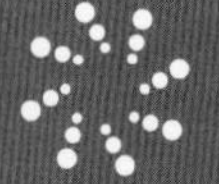

9E.1 Why metals are useful

You should already know | Outcomes | Keywords

Why are metals useful?

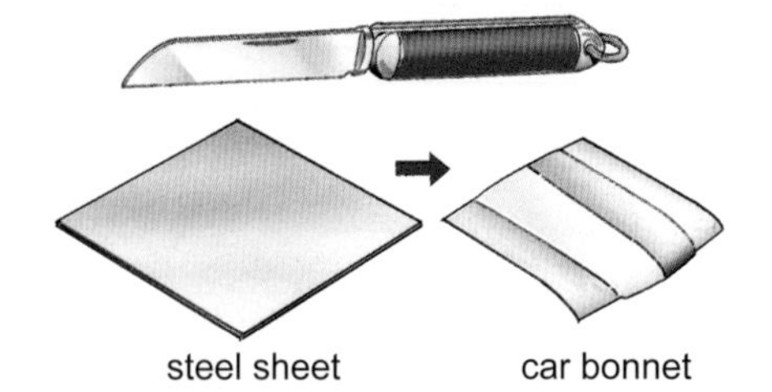

Steel is hard and can be shaped into a knife or a car bonnet.

Metals are used for many different things.

Iron is hard and strong. It is used for making tools.

Iron can be mixed with carbon and small quantities of other metals to make an **alloy** called steel. Steel can be made into different shapes by pressing. This means it is useful in making bodies for cars.

Most types of pan used for cooking are made from metal because metals have high melting points and conduct heat well.

Being able to be shaped, conducting heat, being hard and strong and having a high melting point are all examples of the **properties** of a metal.

The properties of metals make them very useful.

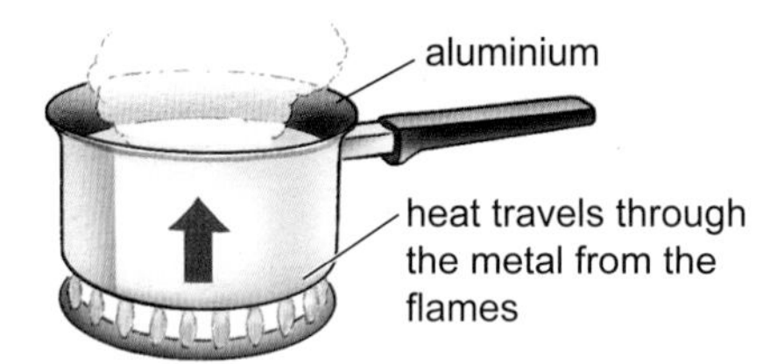

Heat travels through a metal pan.

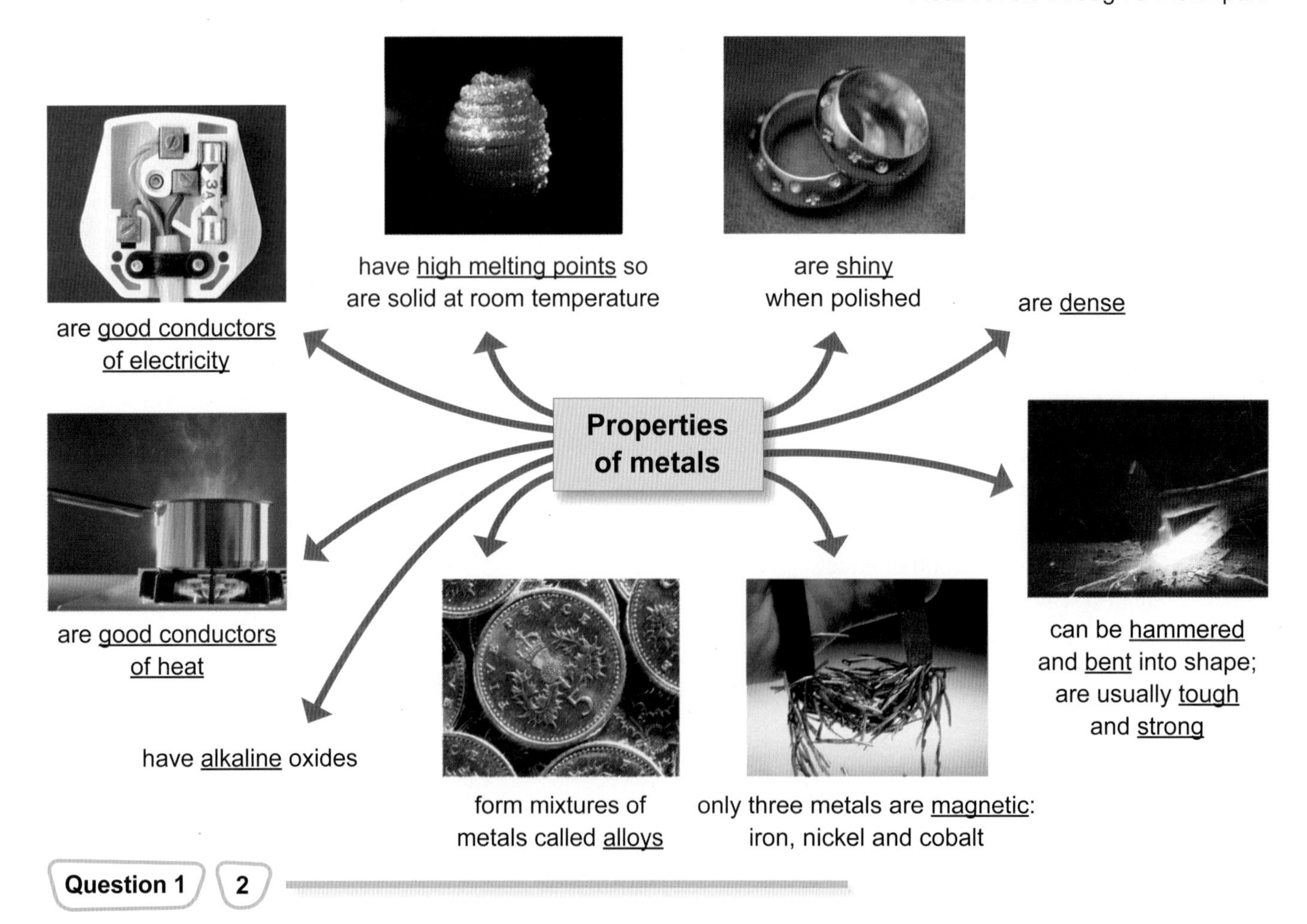

Question 1 | 2

Properties of non-metals

Most of the elements are metals. About a quarter of the elements we find naturally on the Earth are not metals. We call them **non-metals**.

Non-metals are very different from each other. At room temperature some are solids, some are gases and one of them is a liquid.

Even though the properties of non-metals are quite varied, you can think of many of them as being opposite to metals. The table shows you how.

Properties of metals	Properties of non-metals
high melting and boiling points	low melting and boiling points
shiny	dull
can be hammered and bent into shape	brittle when solid
good conductors of heat	poor conductors of heat (are insulators)
form alloys	do not form alloys
good conductors of electricity	poor conductors of electricity (are insulators)

One useful rule is that if an element is a gas at room temperature then it is a non-metal.

Bromine is a liquid that gives off a brown vapour. It is a non-metal.

Chlorine is a green gas. It is a non-metal.

But

There are three famous exceptions to the ideas in the table.

- Mercury is a metal but its melting point is –39 °C, which means it is a liquid at room temperature.
- Graphite is the type of carbon used in pencil leads. It is a good conductor of electricity.
- Diamond is another type of carbon and it is the hardest known natural substance, harder than steel. The scale of hardness goes from 1 to 10. On that scale a steel knife is about 6.5 and diamond is 10!

These unusual properties are put to good use.

- Mercury is used in thermometers.
- Graphite is used in some electrical circuits. It is used in electric motors and generators to make a contact which slides on parts that spin.
- Diamond powder is used on the tips of steel drills that are designed to drill into hard materials.

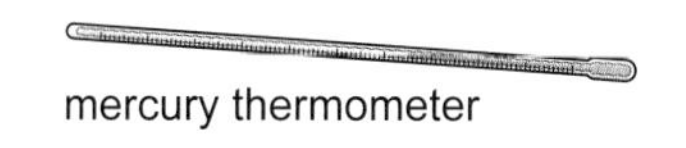

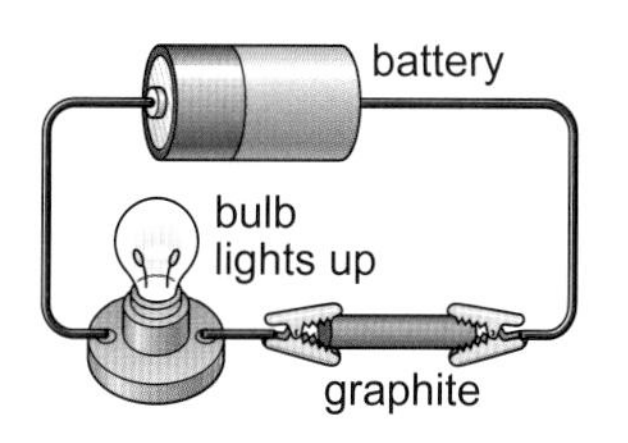

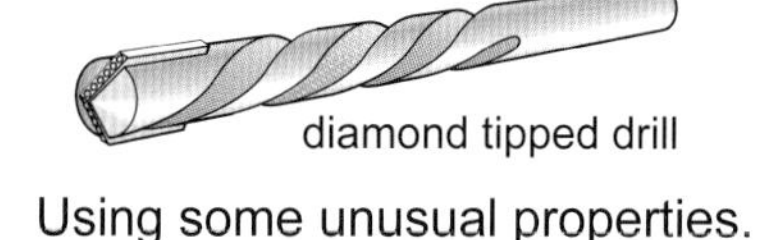

Using some unusual properties.

Question 3  4 5

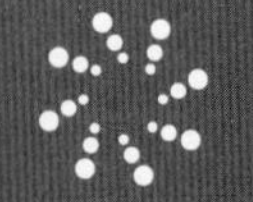

9E.2 Reacting metals with acids

You should already know | Outcomes | Keywords

Metals and acids

Some metals and some acids produce a **chemical reaction** when you put them together.

When zinc reacts with sulfuric acid one of the new substances produced is hydrogen gas. The bubbles of hydrogen gas are easy to see. The other new substance produced is a **salt** called zinc sulfate. This is harder to detect because it is soluble, so it forms part of the solution in the test tube and you cannot see it.

The diagram shows one way to filter off the left-over zinc when the reaction has finished. You get the zinc sulfate from the solution by evaporating the water.

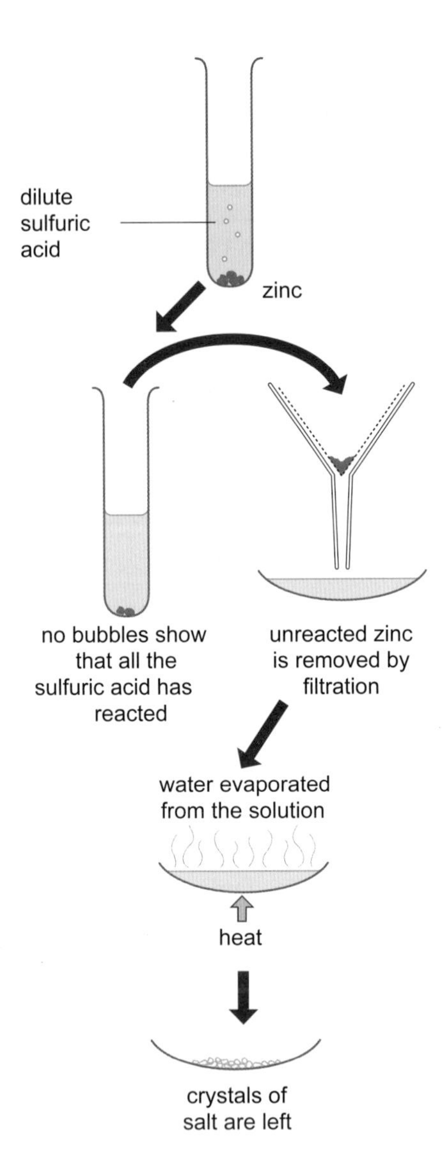

Getting the salt.

You can describe this in a word equation:

zinc + sulfuric acid → hydrogen + zinc sulfate

You can also write this using chemical symbols from the table.

Name	Symbol
zinc	Zn
hydrogen	H
sulfuric acid	H_2SO_4
zinc sulfate	$ZnSO_4$

The small number after the symbol tells you the number of atoms involved. If you have two hydrogen atoms then you write it as H_2.

$$Zn + H_2SO_4 \rightarrow H_2 + ZnSO_4$$

The symbol equation shows you what is happening to the atoms during the reaction. Each Zn atom joins to a combination of a sulfur atom and four oxygen atoms. This leaves the two hydrogen atoms to be released as a molecule of hydrogen gas.

Question 1  3

The name of the salt

The salt produced when zinc and sulfuric acid react is called zinc sulfate.

If a metal reacts with sulfuric acid to make a salt then the salt has 'sulfate' as the second part of its name. The table gives some examples.

Reaction	Salt produced
magnesium and sulfuric acid	magnesium sulfate
aluminium and sulfuric acid	aluminium sulfate

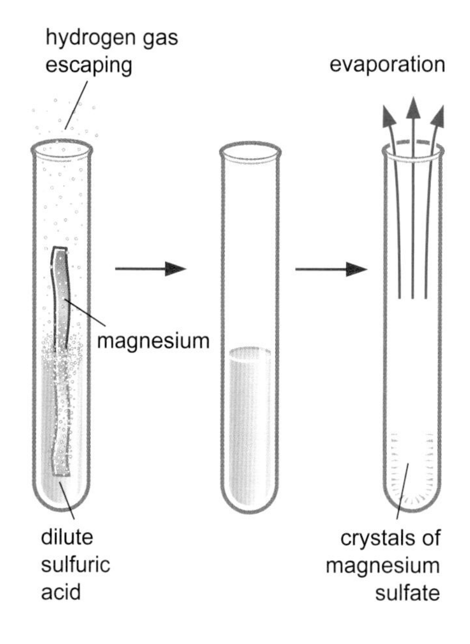

When zinc reacts with hydrochloric acid you get a reaction that is similar to the one with sulfuric acid. Hydrogen gas is produced but this time the salt is different. It is called zinc chloride.

Compare the two word equations:

zinc + sulfuric acid → hydrogen + zinc sulfate

zinc + hydrochloric acid → hydrogen + zinc chloride

The symbol equation for the reaction with hydrochloric acid is

$$Zn + 2HCl \rightarrow H_2 + ZnCl_2$$

- Two hydrogen atoms join together to make one particle of hydrogen gas. This particle is called a molecule of hydrogen. It is written as H_2 because it is made from two atoms of hydrogen.
- A particle of hydrochloric acid has only one hydrogen atom in it so we need two of them to make each H_2 molecule. Two hydrochloric acid particles are written as 2HCl.
- Zinc chloride contains one zinc atom joined to two chlorine atoms. This is shown as $ZnCl_2$. There is only one zinc atom in the products so you only need one zinc atom as part of the reactants. This is shown as Zn with no number in front of it.

zinc reacting with…

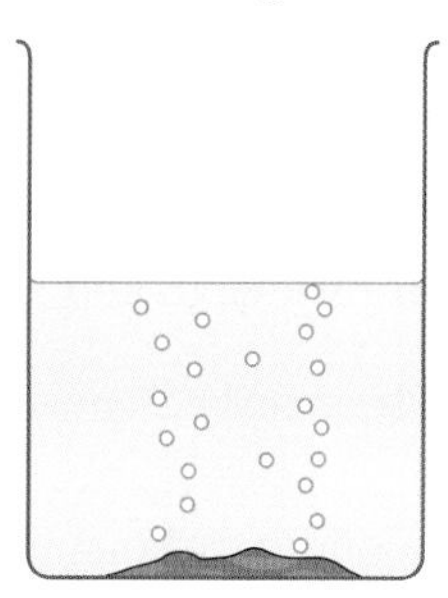

hydrochloric acid

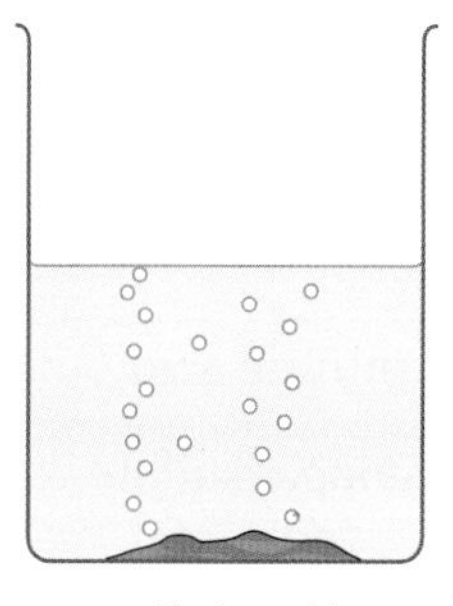

sulfuric acid

The reactions between zinc and sulfuric acid, and between zinc and hydrochloric acid look the same, but they produce different salts.

Not all metals and acids react like zinc does with hydrochloric acid. But if they do, the first part of the name comes from the metal and the second part of the name comes from the acid. This is shown in the table.

Name of acid	Second part of the name of the salt
sulfuric acid	sulfate
hydrochloric acid	chloride
nitric acid	nitrate

Question 4  

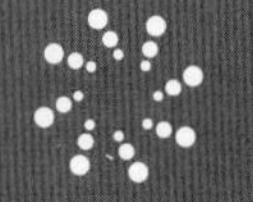

9E.3 Reacting metal compounds with acids

You should already know | Outcomes | Keywords

Metal carbonates

Some combinations of atoms are very common in substances. The combination of one carbon atom and three oxygen atoms is often found joined to a metal atom. The group of one carbon and three oxygen atoms, CO_3, is called a carbonate group. A metal combined with this is called a **metal carbonate**. Examples of metal carbonates include copper carbonate, zinc carbonate, calcium carbonate and magnesium carbonate.

Marble is a rock made from a metal carbonate, calcium carbonate.

If you react a metal carbonate with an acid you get a reaction that follows this pattern.

metal carbonate + acid → salt + water + carbon dioxide

The diagram shows the reaction between copper carbonate and dilute sulfuric acid and how to get the salt out of the solution left at the end.

The word equation for the reaction shown in the diagram is:

copper carbonate + sulfuric acid → copper sulfate + water + carbon dioxide

The formulae for the different substances in this equation are:

Name of substance	Formula
copper carbonate	$CuCO_3$
sulfuric acid	H_2SO_4
copper sulfate	$CuSO_4$
water	H_2O
carbon dioxide	CO_2

This means the symbol equation for the reaction given above is:

$$CuCO_3 + H_2SO_4 \rightarrow CuSO_4 + H_2O + CO_2$$

Question 1  2  3

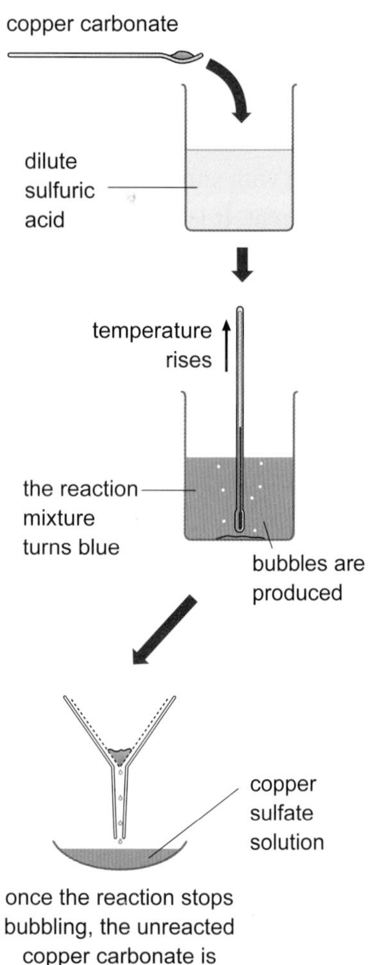

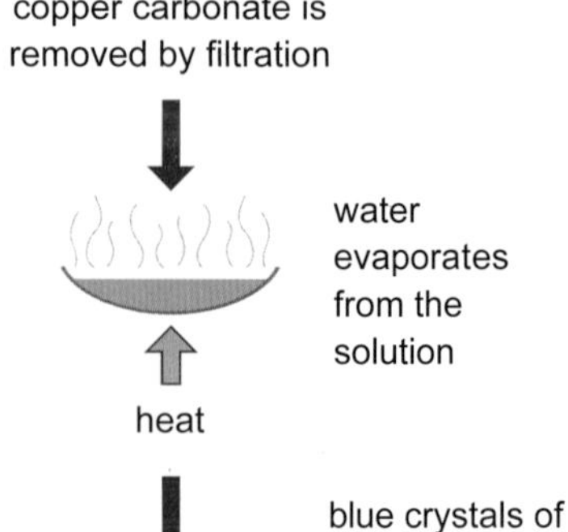

Copper carbonate and dilute sulfuric acid react to form copper sulfate.

Metal oxides

Compounds containing a metal combined with oxygen are called **metal oxides**. These compounds are very common. Many of these compounds are found in the Earth and we get metals from them. We use compounds called **metal ores** in order to manufacture pure metals. Many metal ores consist of metal oxides. Three examples of metal oxides are shown in the table.

Metal oxide	Formula
copper oxide	CuO
sodium oxide	Na_2O
zinc oxide	ZnO

If you react a metal oxide with an acid you always get the same thing happening.

metal oxide + acid → salt + water

The diagram shows how the reaction between copper oxide and sulfuric acid can be used to make the salt, copper sulfate.
The word equation for the reaction is:

copper oxide + sulfuric acid → copper sulfate + water

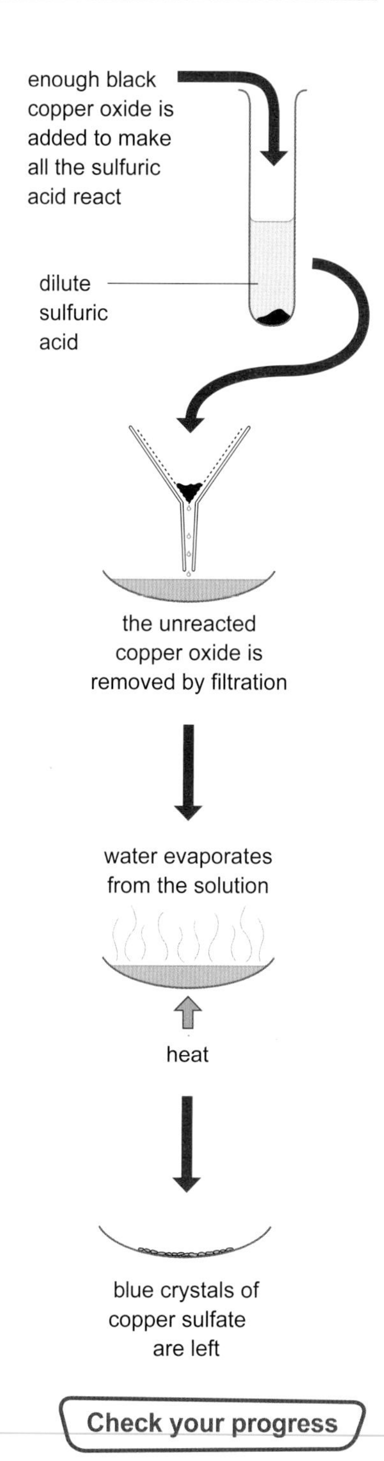

The formulae for the different substances in this equation are:

Name of substance	Formula
copper oxide	CuO
sulfuric acid	H_2SO_4
copper sulfate	$CuSO_4$
water	H_2O

Using the formulae from the table to replace the names gives the symbol equation:

$$CuO + H_2SO_4 \rightarrow CuSO_4 + H_2O$$

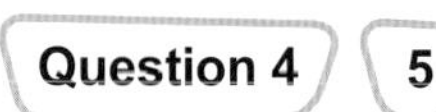

The name of the salt

The name of the salt formed when a metal carbonate or a metal oxide reacts with an acid follows the same idea as the name of the salt formed when a metal reacts with an acid.

- The first part of the name comes from the metal.
- The second part of the name comes from the acid.

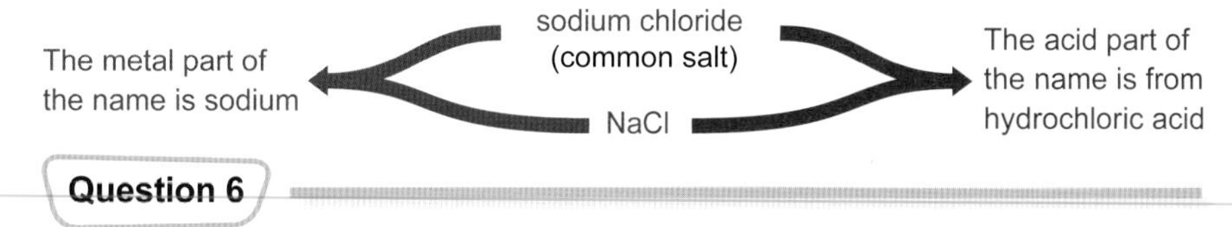

Question 6

Check your progress

9E.4 Neutralisation and salts

You should already know

Outcomes

Keywords

Types of reaction

Calcium carbonate neutralises the effect of acid rain.

You have already met some types of reaction.

- **Combustion reactions** involve a substance burning to combine with oxygen.
- **Thermal decomposition reactions** are when a substance splits up into new substances when it is heated.

The reactions in this Unit are examples of a different type of reaction, called a **neutralisation reaction**. Examples include:

- metal carbonate + acid;
- metal oxide + acid.

In a neutralisation reaction, an acid reacts with a substance that neutralises the acid to form a salt.

Neutralisation reactions are used in the everyday world to reduce or eliminate the effect of acids.

When a bee stings, it injects an acid into the skin. Calamine lotion can be used to treat the sting because calamine lotion contains zinc carbonate, which will neutralise the acid.

Baking powder can be used to treat a bee sting as an alternative to calamine lotion. Baking powder is made from sodium hydrogencarbonate. This is a substance that will neutralise the acid.

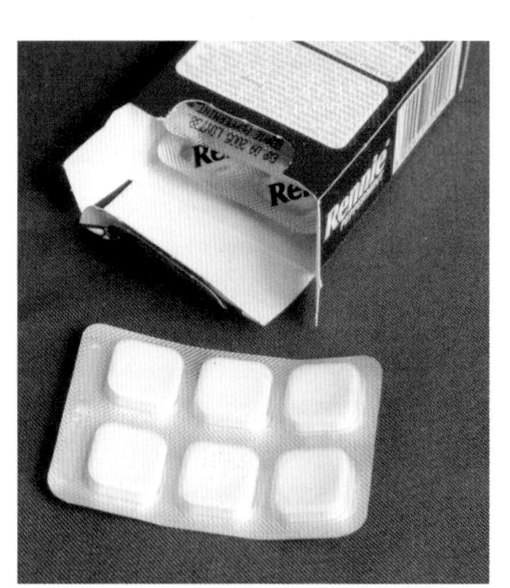

Indigestion tablets neutralise excess stomach acid.

When a wasp stings, it injects an alkali so you cannot use a metal carbonate to treat the sting. You can treat a wasp sting with vinegar because vinegar is an acid and neutralises the alkali from the sting.

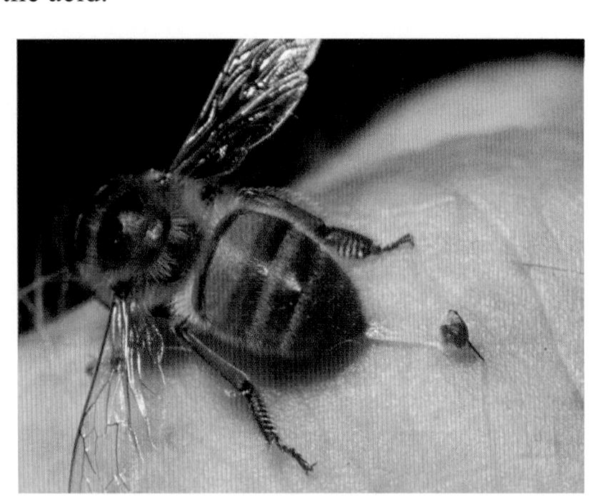

Calamine lotion neutralises the acid in bee stings.

Question 1  2

Uses of salts

Salts are important substances. They have many different uses. Some examples are given in the photographs.

Gunpowder contains potassium nitrate.

Copper sulfate, used as a fungicide in agriculture.

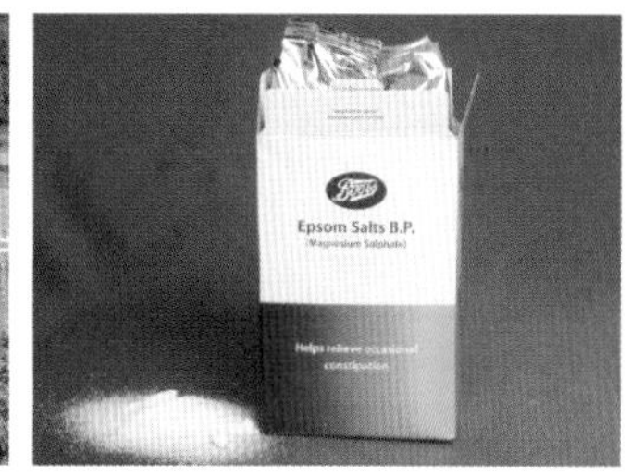

Epsom salts are magnesium sulfate.

Photographic film contains silver nitrate.

Question 3

Sodium chloride is the salt used in cooking and is often called 'common salt'.

You can make common salt by neutralising hydrochloric acid with the alkali sodium hydroxide. The diagram shows how.

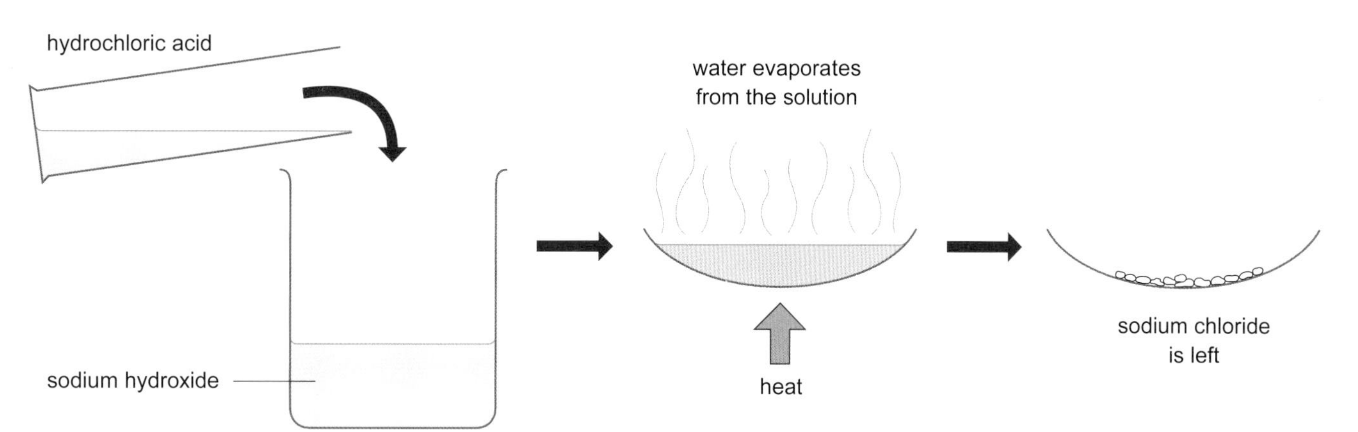

You have to add exactly the right amount of acid to the alkali to get a neutral solution of salt in water. This can be quite difficult to do unless you use special equipment and an indicator like universal indicator.

Common salt is not produced commercially by this method.

Common salt is usually extracted from sea water by letting the sea water evaporate in the sun. You can also dig salt out of the ground in a salt mine. Salt dug out of the ground is called rock salt. Rock salt has to be purified unless it is only going to be used directly on roads in icy weather.

A modern salt mine.

Question 4

Review your work

Summary ➡

You should already know | Outcomes | Keywords

Neutralisation

Neutralisation reactions between an acid and an alkali can be used to make a neutral salt. The problem is how to tell when the exact amount of alkali needed to get to the point of neutralisation has been added to the acid.

One way to approach this problem is to use universal indicator. A neutral solution has a pH of 7. Universal indicator is green when the pH of the solution is 7.

You can mix an alkali and an acid and use the colour of the universal indicator to check the value of the pH. The pH of the solution is a **quantitative** result because it is a number and not just a description.

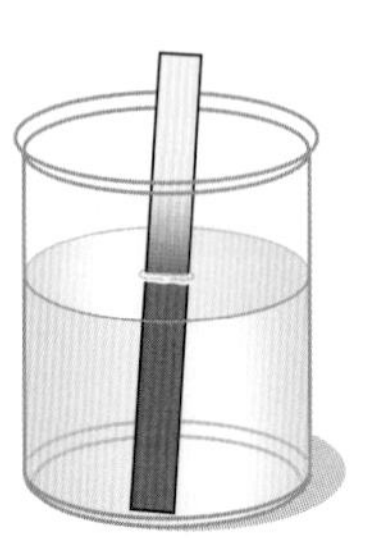

dilute hydrochloric acid, strongly acidic, pH = 1

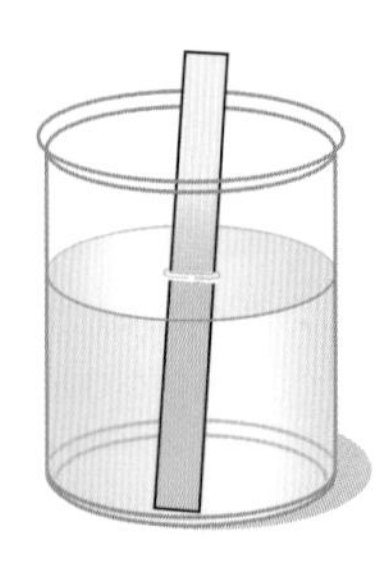

neutral solution, pH = 7

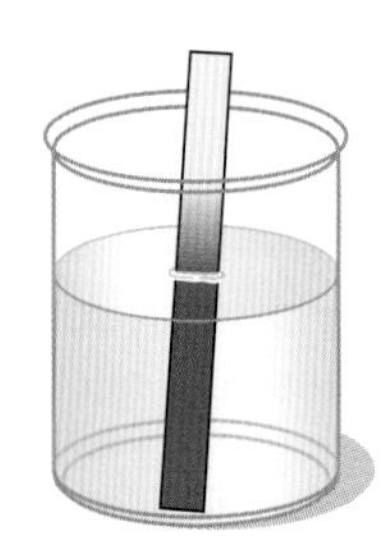

dilute potassium hydroxide, strongly alkaline, pH = 12

red	orange	yellow	green	blue	navy blue	purple
0–2	3–4	5–6	7	8–9	10–12	13–14

more acidic ← neutral → more alkaline

pH and universal indicator.

Question 1 

A simple method

A simple method of adding the alkali to the acid is to add drops of alkali to a test tube containing acid. You can get a quantitative result for the amount of alkali you add by counting the drops.

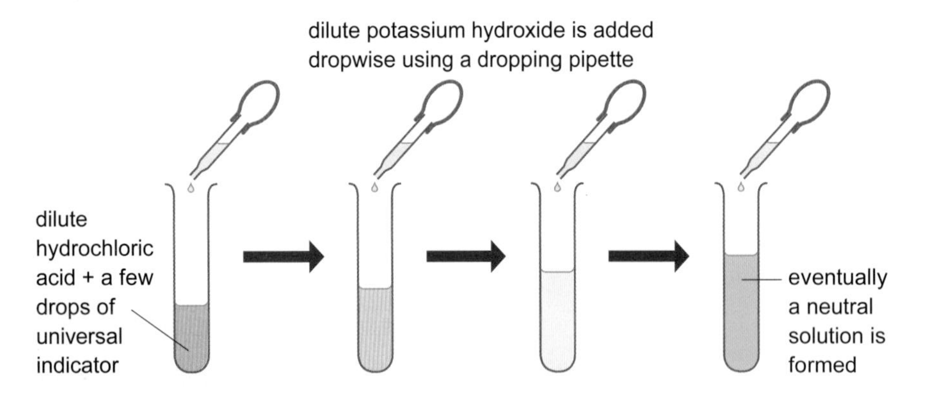

You can get a more accurate quantitative result by using a **burette**. This is a tube with a tap. It has an accurate volume scale, usually calibrated in **millilitres**. The tap allows the alkali to be added in very tiny amounts.

Question 3

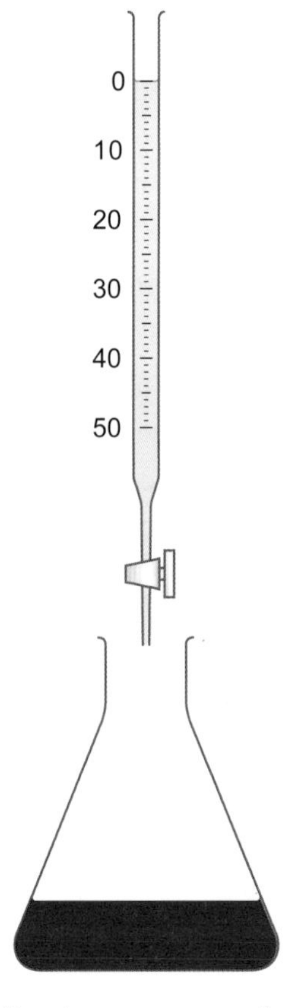

A burette gives an accurate quantitative result.

Another approach to measuring the pH

One problem with the previous method is the use of universal indicator which is added to the acid. This means it is in the solution at the end of the experiment. Although you want to know when the pH is 7, you might not want any additional substances in the salt solution.

Measuring the pH with universal indicator depends on how good you are at judging a colour. The judgement of colour can depend upon the person doing the judging. We say it is **subjective**. This is made even worse if the observer suffers from some form of colour blindness.

One way round these problems is to use a meter attached to a sensor called a pH probe. This does not contaminate the solution. It also shows the number for the pH on a digital display. This means that the reading does not depend upon the person looking at it. We say the result is **objective**. Objective evidence is much more reliable than subjective evidence.

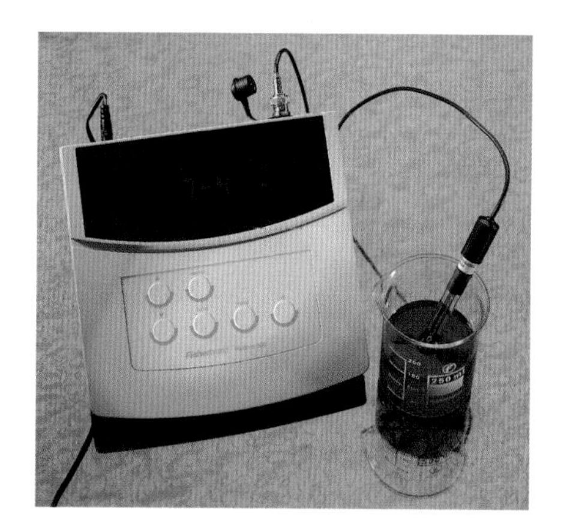

Using a pH probe.

Question 4 **5**

Another approach to finding the volume

If you add the alkali to the acid 1 cm^3 at a time from a burette, give the solution a good stir and measure the pH each time with the pH probe, you will get results like these.

Volume of potassium hydroxide added (cm^3)	0	1	2	3	4	5	6	7	8	9
pH of solution	1	1	1	2	3	10	11	12	12	12

These results are very accurate but they do not show you straight away how much alkali is needed to neutralise the acid.

You can plot a graph to show the results more clearly.

The graph shows how the pH of the solution changes as the volume of alkali increases. The neutral solution is formed when the pH is 7.

A line is drawn across from pH 7 on the vertical axis until it meets the line of the graph. A line is drawn from this point on the graph down to the horizontal axis. This tells us how much alkali we need to get to a pH of 7.

We can see from the graph that this happens when 4.6 cm^3 of alkali has been added to the acid. The graph makes it easy to see how the pH changes as you add the alkali to the acid.

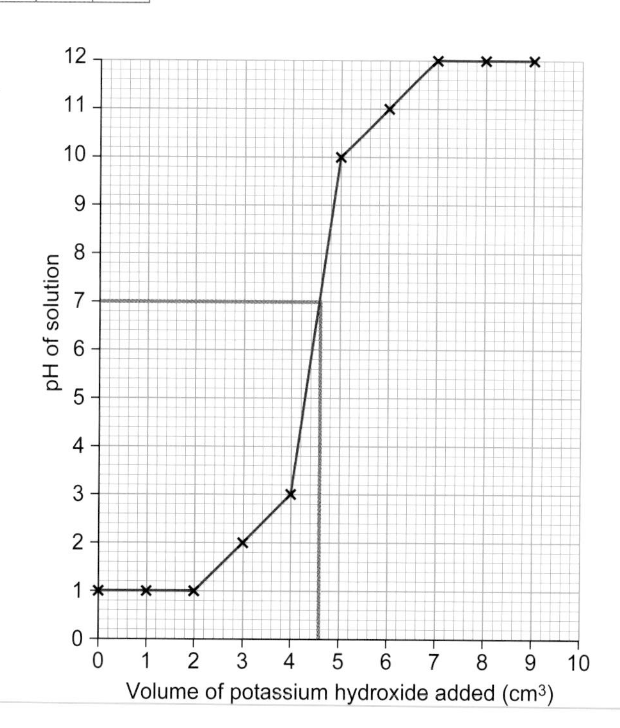

Question 6  **7**

9E Questions

9E.1

1 Give two properties of iron that make it useful for making tools.

2 a Write down six properties that all metals have in common.

b What property do the three metals cobalt, nickel and iron have in common that other metals do not have?

3 Give six typical properties of non-metals.

4 An element is a gas at room temperature. Is it a metal or a non-metal? Give a reason for your answer.

5 Give three examples of elements that have unusual properties. State what these properties are and how they can be put to use.

9E.2

1 What new substances are formed when zinc reacts with sulfuric acid?

2 Write down the word equation for the reaction between zinc and sulfuric acid.

3 Write down the chemical symbol for one particle of zinc sulfate and explain what the symbol tells us.

4 If a metal reacts with sulfuric acid, what is the second part of the name of the salt that is produced?

5 Write down the symbol equation for this reaction:

zinc + hydrochloric acid → hydrogen + zinc chloride

6 What acid could be used to make a salt with 'chloride' in its name?

9E.3

1 Name the elements found in:
  a calcium carbonate;
  b sulfuric acid.
2 Write down the word equation for the reaction between a metal carbonate and an acid.
3 a Write down the word equation for the reaction between copper carbonate and sulfuric acid.
  b Write down the symbol equation for the same reaction.
4 a What elements are found in a metal oxide?
  b Give two examples of metal oxides.
5 Write down the word and symbol equations for the reaction between copper oxide and sulfuric acid.

9E.4

1 Give two different examples of a neutralisation reaction.
2 Explain how neutralisation reactions can be used to treat wasp and bee stings.
3 Give four examples of different salts and their uses.
4 Describe two ways of making common salt, one in the laboratory and one that is used for commercial production.

9E.HSW

1 How can universal indicator be used to tell when an acid has neutralised an alkali?
2 a What is meant by quantitative results?
  b Give one example.
3 What is the advantage of using a burette over counting drops as a quantitative method?
4 a What is meant by a subjective result?
  b Give one example.
5 What is the advantage of using a pH probe over using universal indicator?
6 What is the advantage of using a graph over putting the results in a table?
7 Use the graph to work out what volume of potassium hydroxide would be needed to get a pH of 11.

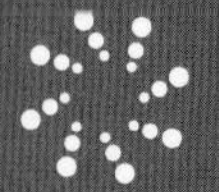

Scientific enquiry

9F.1 Metals and water

You should already know | Outcomes | Keywords

Metals and water

Some **metals** do not react with water. This makes them useful. Some examples are given in the table.

Metal	Comment
Copper	Copper is a very good heat conductor and does not react with water. It is used for water pipes, particularly in heating systems, and for high quality cooking pans.
Silver	Silver is an attractive shiny metal that does not react with water. Expensive cutlery and wine goblets are sometimes made of silver. You also get cheaper cutlery made from other metal but coated with a layer of silver.
Gold	Gold does not react with water. If it did, you would have to take off any gold rings before washing your hands.
Stainless steel	This metal is a mixture of other metals called an alloy. As well as iron and other metals like chromium, it also contains the non-metal carbon. Steel does not react with water. It is used for cooking pans and cutlery.

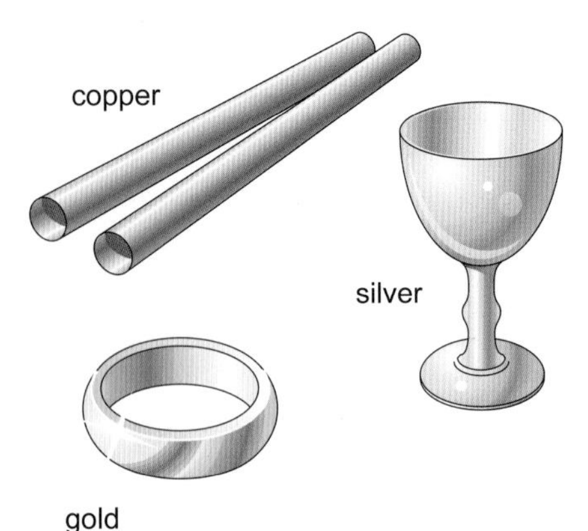

Three different metals.

A high quality chef's pan made from copper.

Question 1 | 2

Magnesium and water

Magnesium is a metal that does react with water. The **reaction** is very slow. If you put a strip of magnesium into cold water you have to leave it for quite a while before you notice the reaction. Eventually tiny bubbles of hydrogen appear on the surface of the magnesium. The picture shows the effect.

The reaction is:

magnesium + water → magnesium oxide + hydrogen

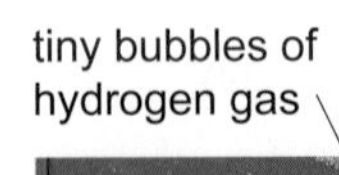

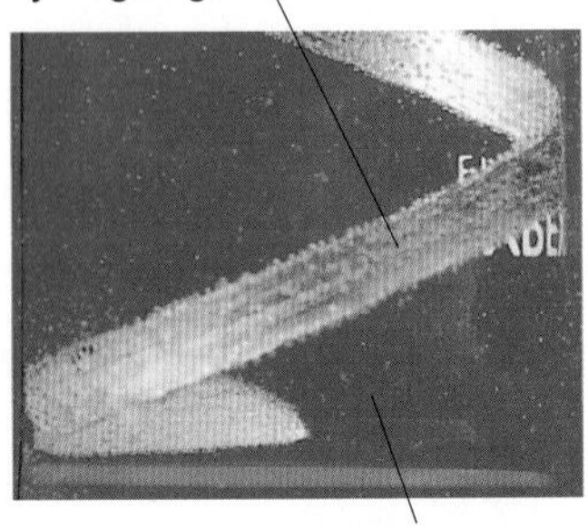

Magnesium reacts very slowly with cold water.

Question 3

Magnesium and steam

The reaction between magnesium and steam is much more spectacular. Magnesium actually burns in steam!

Magnesium will react very slowly with water, but more quickly when the water has turned to steam. Particles in steam have a lot more energy so the reaction happens a lot more easily.

The symbol equation is: $Mg + H_2O \rightarrow MgO + H_2$

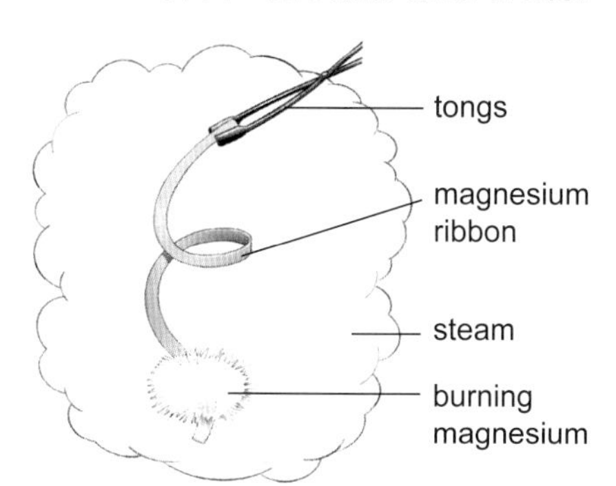

Magnesium burns in steam. Hydrogen and magnesium oxide are the two new substances made in this chemical reaction.

Question 4

Sodium and water

Sodium is very **reactive** with water. It will even react with the water vapour in the air so it has to be stored in a jar of oil.

Sodium is stored under oil.

If you put a piece of sodium into water it fizzes about on the surface producing bubbles of hydrogen.

The word equation for the reaction is:

sodium + water → sodium hydroxide + hydrogen

The metals that are in the same group as sodium in the periodic table are lithium, potassium, rubidium, caesium and francium. This group is known as Group 1 in the periodic table. All metals in Group 1 have the same type of reaction with water but with rubidium the reaction is very dangerous and violent. Even with potassium, the reaction produces enough heat to set fire to the hydrogen produced.

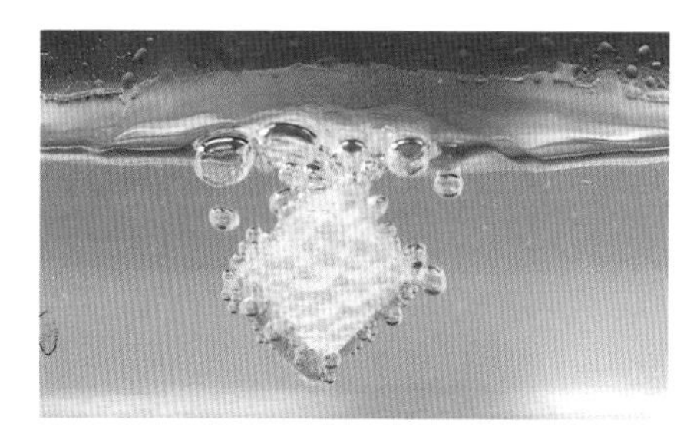

sodium + water → sodium hydroxide + hydrogen

The reaction produces an alkali. In the case of sodium, the alkali is sodium hydroxide. That is why the group of metals is known as the **alkali metals**.

A range of reactivity

Some metals do not seem to react with water but they do, it just takes a long time. Examples of these metals are iron and zinc.

Iron is such a useful metal that we spend a lot of time and money protecting it so it is not exposed to water. A good example of this is the paint used on the bodies of cars.

The word used for how fast something reacts is 'reactivity'. The list shows how fast some metals react with water. The most reactive metals are shown at the top. This type of list is based on the reactivity of the metals with water. It is called a **reactivity series**.

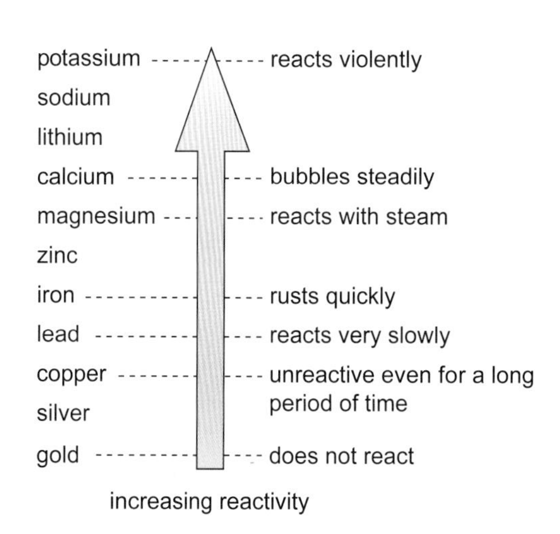

Reactivity series for metals reacting with water.

Question 5  **6** **7**

9F.2 More about metal reactions

You should already know

Outcomes

Keywords

Metals and acid

Only a few metals react quickly with water. Many more metals react quickly with hydrochloric acid. When a metal reacts with hydrochloric acid it is usually easy to see that a reaction is happening. Bubbles of hydrogen gas appear and the metal gradually disappears. The other new substance produced is a **salt** with '**chloride**' as the last part of its name.

The diagram shows bubbles of hydrogen being produced when magnesium is added to hydrochloric acid. The solution formed as the metal is eaten away contains magnesium chloride.

The diagram also shows the effect of temperature. The reaction is much quicker in hot acid than it is in cold because the particles move more quickly at higher temperatures.

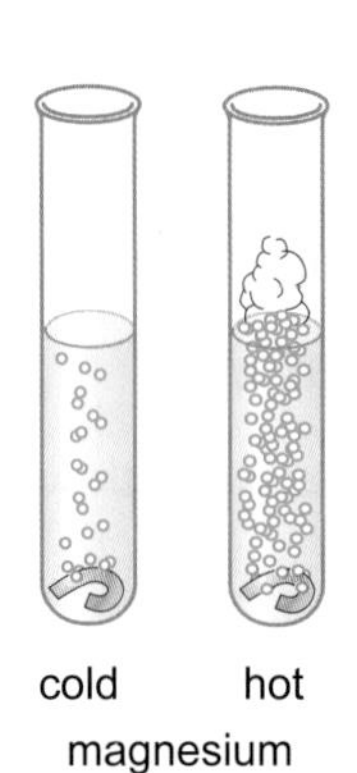

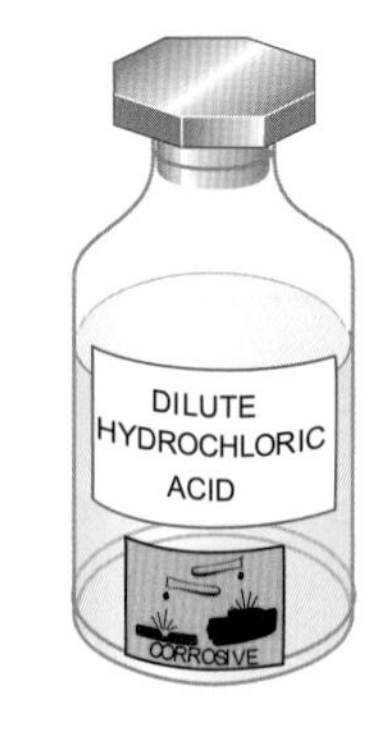

Question 1 | 2

More about the speed of reaction

The reaction between a metal and an acid is faster if the temperature is higher. It is also faster if the concentration of the acid is higher. In very dilute hydrochloric acid, it might appear that metals such as iron and aluminium do not react. If you heat the acid up you can see they do react.

cold hot
aluminium

cold hot
iron

Copper and gold do not react with hydrochloric acid at all.

Metal	Reaction with hydrochloric acid
magnesium	vigorous reaction
zinc	quite slow
iron	slow
lead	has a slow reaction if the acid is concentrated
copper and gold	no reaction

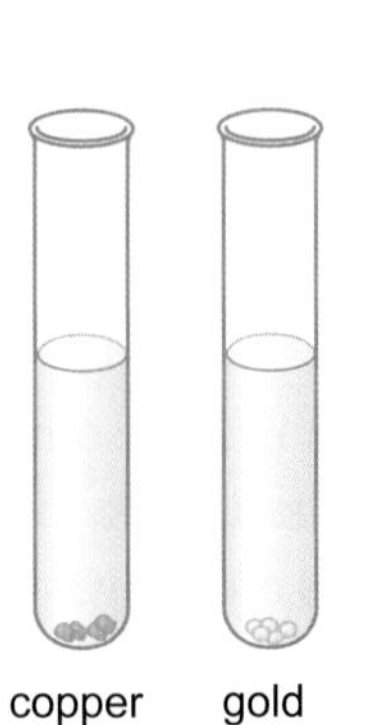

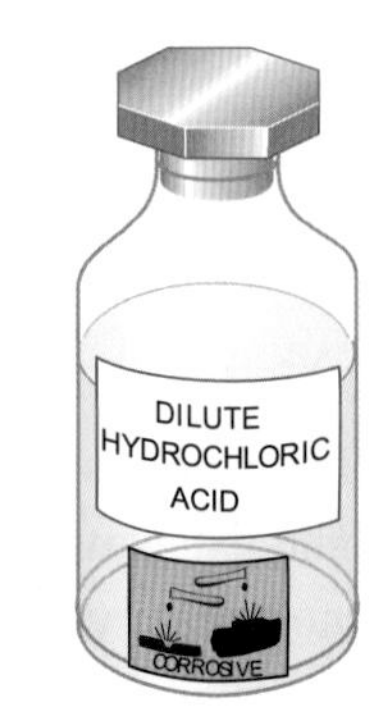

Question 3

Metals reacting with oxygen

About one fifth of the atmosphere is oxygen. Many metals react with oxygen, particularly if they get hot.

You can see a spectacular effect of this in the sky at certain times of the year when the Earth passes through a shower of meteorites. Some meteorites are made of iron. They burn brightly in the oxygen in the atmosphere. We call them shooting stars.

Shooting stars fall to the ground as meteorites.

iron + oxygen → iron oxide

The bigger meteorites do not burn up completely and they land on the ground. We can examine them to see what they are made from.

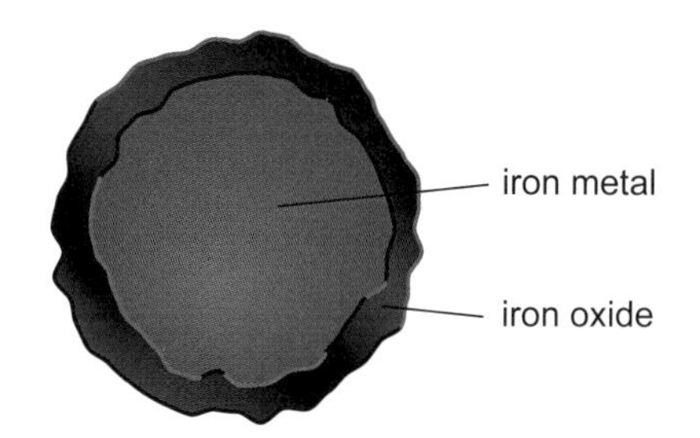

A meteorite cut in half (cross-section).

The same idea is used in a sparkler. The sparks are bits of hot iron burning as they fly through the air. The word equation is:

iron + oxygen → iron oxide

Iron also reacts with a combination of oxygen and water to make a substance we call 'rust'. Rust is a lot weaker than iron so it is important to stop rust forming on things like car bodies, iron tools and iron bridges. Steel is made from iron and it also needs protecting from oxygen unless it is specially made so it does not rust. This steel is called stainless steel.

The pictures show some ways of protecting iron and steel.

An iron chisel is protected with a film of oil.

Cars are painted.

Food cans are steel with a thin coating of tin.

Painting steel ropes on a suspension bridge.

Other metals react with oxygen as well. For example, magnesium will burn with a bright white flame.

Question 4 5 6 

Putting metals in order

By observing how metals react with water, dilute acid and oxygen, scientists have drawn up a reactivity series for them. A shortened list is shown here.

potassium
sodium
lithium
calcium
magnesium
zinc
iron
lead
copper
silver
gold

increasing reactivity with oxygen

Again, we can make a reactivity series and use it to make predictions about other metals.

You should already know | Outcomes | Keywords

Metals pushing other metals out of the way

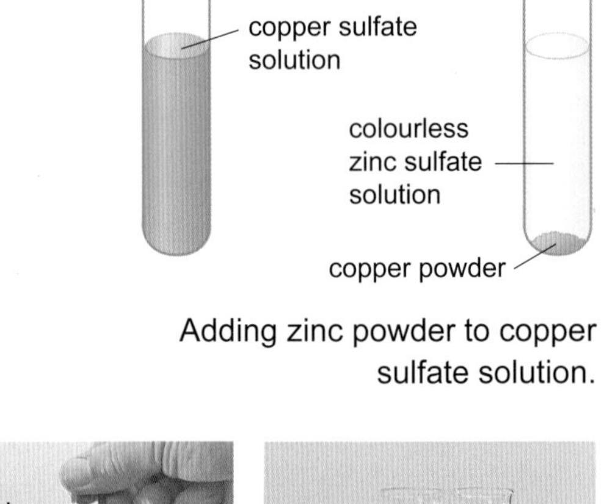

Adding zinc powder to copper sulfate solution.

If you look at the reactivity series you can see that zinc is more reactive than copper. This means that if you put zinc powder into copper sulfate solution the zinc is able to push the copper out of the solution.

The zinc takes the place of the copper in the solution. The word equation for what happens is:

zinc + copper sulfate → zinc sulfate + copper

The drawing shows what you observe if you do this.

We call this a **displacement reaction** because the zinc has displaced the copper.

A more reactive metal will displace a less reactive metal from a solution. There are many examples. The photographs show magnesium displacing copper from copper sulfate. This happens because magnesium is more reactive than copper, so the magnesium pushes the copper out of the solution. We say the magnesium displaces the copper.

The word equation is:

magnesium + copper sulfate → magnesium sulfate + copper

If you put copper into magnesium sulfate nothing happens. Copper is less reactive than magnesium so it cannot displace it.

Another displacement reaction happens if you dip an iron nail into copper sulfate solution. The iron on the surface of the nail displaces the copper so the outside of the nail becomes a copper layer and the solution changes to iron sulfate.

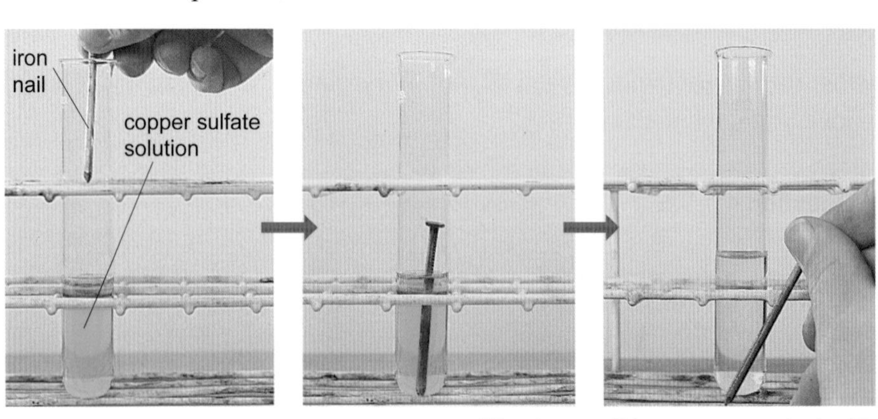

The iron nail becomes coated with copper, and the liquid changes to iron sulfate solution.

Question 1 | 2

Using displacement reactions

Aluminium is more reactive than iron. Aluminium will displace iron from the solid compound iron oxide if you heat it up.

The word equation for the reaction is:

aluminium + iron oxide → aluminium oxide + iron

This reaction also produces a large amount of energy so the iron is produced as molten iron at a very high temperature. The melting point of iron is 1535 °C, which is very hot indeed. Compare that to a pizza oven at about 300 °C!

The photograph shows the reaction being used to weld railway lines together. The reaction happens in the can and the molten iron pours out of the bottom into a mould on the railway lines.

Using carbon

Carbon is not a metal but you can use it to displace some metals and not others. Carbon will displace zinc, iron, tin and lead from their ores.

This was discovered about 3500 years ago when people discovered you could get iron from **iron ore** by heating it to a very high temperature with charcoal, which is a form of carbon.

We get our iron today in a similar way on a large scale in a container called a blast furnace.

How they made iron in the Iron Age.

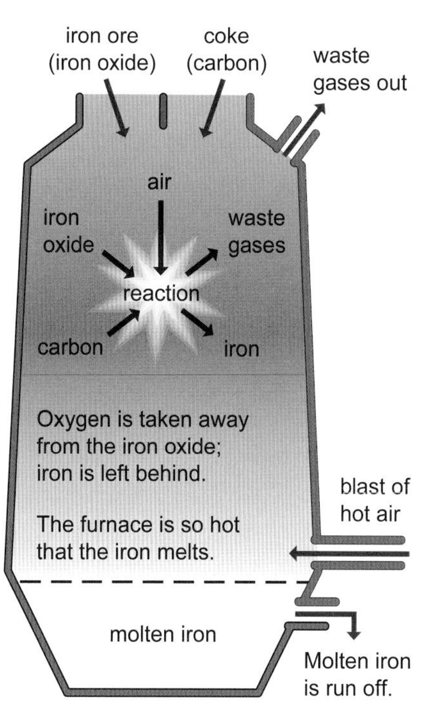

How we get iron from iron ore today in a blast furnace.

Check your progress

9F.4 More ideas about metals

You should already know

Outcomes

Keywords

Metals in the Earth

Gold and silver are not very reactive. They can be found as small lumps of metal in the Earth's crust.

Native gold.

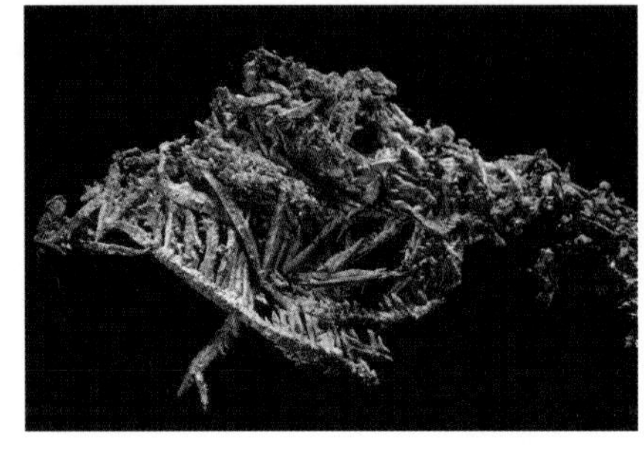

Native silver.

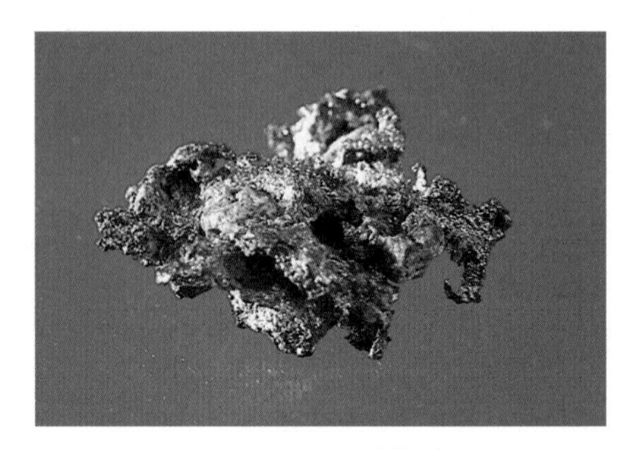

Native copper.

Metals that are found as pure metal in the ground are called **native metals**. Sometimes copper can also be found as a native metal. Gold, silver and copper are three of the few metals that can be found as native metals. Most metals cannot be found as native metals. They are found in compounds in rocks. The rocks we use to get metals from are called **ores**.

Rock salt is mostly sodium combined with chlorine.

Bauxite is an ore that we get aluminium from. It is mainly aluminium combined with oxygen.

Galena is an ore that we get lead from. It is mainly lead combined with sulfur.

Haematite is an ore that we get iron from. It is mainly iron combined with oxygen.

Extracting metal from the ore

Iron, lead, zinc and tin can be extracted from their ores by displacing them with **carbon**. This is because carbon is more reactive than these metals. Metals like aluminium are more reactive than carbon, so carbon cannot be used to extract them from their ores. In the 19th century, scientists discovered how to extract metals that are more reactive than carbon by using electricity. This is the method we use today to get the metals aluminium, calcium, barium and magnesium.

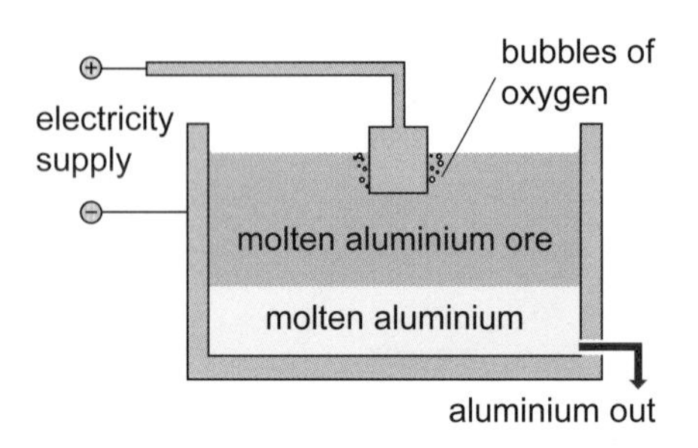

We extract aluminium by melting aluminium oxide and passing electricity through it.

Question 1 2

Alloys

An **alloy** is a mixture of a metal and another substance. Alloys have different properties to the metals that they are made from.

Solder is an alloy. Solder is used to join metals together in electrical circuits. It is usually made from tin and lead. It melts more easily than either lead or tin.

Bronze is an alloy made from tin and copper. It is very hard. It does not corrode easily. It is used for statues made of metal and for church bells.

Alloy	Metals in it
bronze	copper and tin
brass	copper and zinc
pewter	lead and tin
steel	iron and carbon

Question 3 

Dates of extraction

In 1800 only about 12 metals were in common use. In about 1810, people discovered that reactive metals could be extracted from ores using electricity. By 1899 another 41 metals had been extracted from their ores. The availability of metals helped the Industrial Revolution happen in Britain. This produced a massive increase in technology and changed the way most people lived in Britain.

Metal	Date of first extraction	Example of modern use
gold	6000 BC	jewellery
copper	4200 BC	electrical wiring
iron	1500 BC	railway lines
sodium	AD 1807	street lighting
aluminium	AD 1827	overhead cables

Question 5

Review your work

Summary ➡

9F.HSW Investigating how different metals react

You should already know

Outcomes

Keywords

Investigating how different metals react with acid

The diagram shows four different metals in hydrochloric acid.

You can see that:

- copper docs not react;
- magnesium reacts more than zinc and iron;
- zinc and iron react in similar ways.

These are **qualitative** observations. They are descriptive. They do not have any numbers associated with them.

If you took measurements to compare how iron and zinc react with hydrochloric acid, you would be making **quantitative** observations. The experiment would be described as a quantitative experiment.

Question 1 2

Making a quantitative observation

One quantitative experiment is to count the bubbles of hydrogen that come off when a metal reacts with an acid.

You need to make it a **fair test** by:

- using identical size pieces of metal;
- using the same strength of acid each time;
- using the same temperature of acid each time;
- counting for the same length of time.

An important idea is to change only one thing at a time in an experiment, so you know what is affecting the results. When we keep things the same so they do not affect the result, we call them **control variables**. You make something a fair test by controlling all the variables and only changing one of them.

Question 3

magnesium

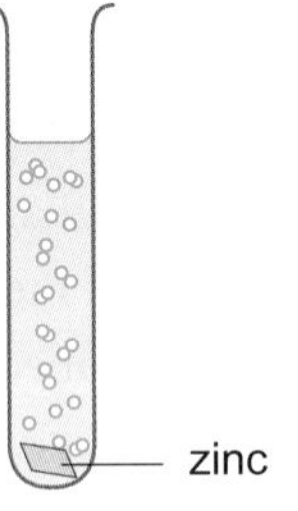

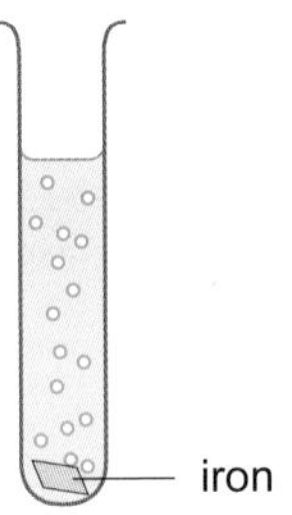

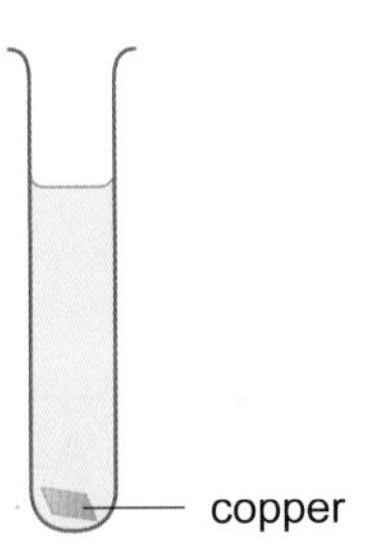

Reacting metals with hydrochloric acid.

Measuring the volume

It is difficult to count the bubbles of gas accurately even if you use a very weak acid. One way to get a quantitative result for this investigation without counting bubbles is to collect the gas that is produced in a burette.

The burette is full of water at the start. Bubbles of gas push the water out of the burette as they are produced.

This method gives a direct result for the volume of gas produced rather than a number of bubbles. The volume of gas produced is shown on the scale on the burette.

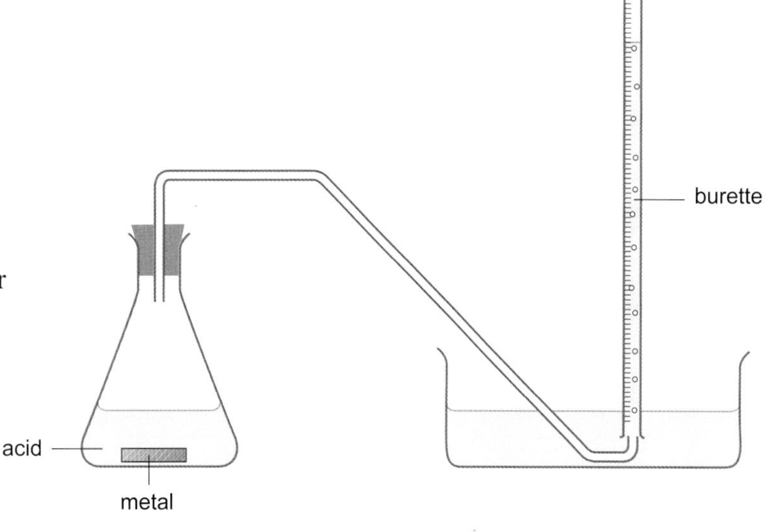

The hydrogen produced pushes the water out of the burette.

Question 5

If you do this experiment you get results like these.

Time in seconds	Volume of gas produced, in cm^3			
	With copper	With magnesium	With iron	With zinc
30	0	4.0	1.0	1.5
60	0	8.0	3.0	3.0
90	0	13.0	5.0	6.0
120	0	18.0	7.0	9.0
150	0	22.5	9.0	12.0
180	0	26.0	10.5	14.0
210	0	28.0	11.5	15.5

After 210 seconds, magnesium has produced more gas than copper, iron or zinc. You can see which metals have reacted more than others by looking at the results for different metals at the same time.

The volume of gas will stop increasing when one of these events happens.

- All the acid has been used up.
- All the metal has been used up.

Question 6

9F.1

1 Give <u>three</u> examples of situations where metals are used because they do not react with water.

2 Suggest why cheaper metals are coated with silver to make some types of cutlery.

3 How can you tell that magnesium reacts with water?

4 Write the word equation for the reaction between magnesium and steam.

5 Name <u>three</u> other metals that react in a similar way to sodium when they are put into water.

6 **a** What is the name for the family of metals that sodium belongs to?

b Why is sodium described as one of the alkali metals?

7 **a** Give <u>two</u> examples of metals that are below copper in the reactivity series.

b Write down <u>one</u> prediction you can make about their reaction with water.

9F.2

1 What are the products if a metal reacts with hydrochloric acid?

2 **a** What is the effect on a reaction between a metal and hydrochloric acid if the acid is heated up?

b Suggest <u>one</u> reason why this is so.

3 **a** What is the difference between the reaction of lead with hydrochloric acid and the reaction of iron with hydrochloric acid?

b How could both these reactions be speeded up?

4 Write down the word equation for the reaction between iron and oxygen.

5 Why is rust a problem?

6 Describe <u>three</u> methods of preventing iron from rusting.

9F.3

1 **a** Why is zinc able to push copper out of the way if you put zinc powder into copper sulfate solution?

b What observation shows that this has happened?

2 **a** Explain why an iron nail dipped into copper sulfate solution becomes coated in copper.

b What happens to the solution?

continued

3 Aluminium will displace iron from iron oxide.
Give one practical use of this.

4 a Why can carbon be used to extract zinc from its ore?
b Name three other metals that can be extracted from their ores using carbon.

5 Describe how a blast furnace is used to extract iron from iron ore.

9F.4

1 a What is meant by a native metal?
b Give three examples and suggest a reason why these metals can be found as native metals.

2 a What method is used to extract less reactive metals like iron from their ores?
b Give one reason why this method can be used.

3 What is solder usually made from and what is it used for?

4 What is bronze made from?

5 What development enabled the extraction of many metals from their ores after 1810?

9F.HSW

1 a What is a qualitative observation?
b Give an example.

2 Why are qualitative observations not very useful for comparing the ways zinc and iron react with hydrochloric acid?

3 List four factors that might affect the results of an experiment to compare how different metals react with hydrochloric acid by counting bubbles.

4 a What is a control variable?
b How are control variables used to make an experiment a fair test?

5 a Explain how the use of a burette can provide a quantitative result in an experiment to measure the volume of gas produced when a metal reacts with an acid.
b Why is this preferable to a method based on counting bubbles?

6 a Plot a graph showing the results for magnesium, iron and zinc on the same set of axes. Put volume of gas on the vertical axis and show time along the horizontal axis.
b What is the advantage of the graph over the table?

Scientific enquiry

9G.1 Soils

You should already know

Outcomes

Keywords

What is in soil?

If you put some soil in a tall beaker and shake it up with a lot of water the mixture will look like the diagram.

Soil consists of a variety of substances. Different proportions of these substances make very different soils:

- grains that have come from the weathering and erosion of rocks, like sand, clay or silt;
- dead plant and animal material that is rotting – this is called **humus**;
- air in the spaces between the grains that make up soil;
- water caught in some of the spaces between the particles;
- living things – some of which you can see, like worms, and many you cannot see, like bacteria;
- small stones and gravel.

When humus breaks down it adds **mineral nutrients** to the soil. The plant growth, called vegetation, is therefore very important because dead leaves that fall to the ground help form humus. Animals that burrow, like worms, help get air into the soil.

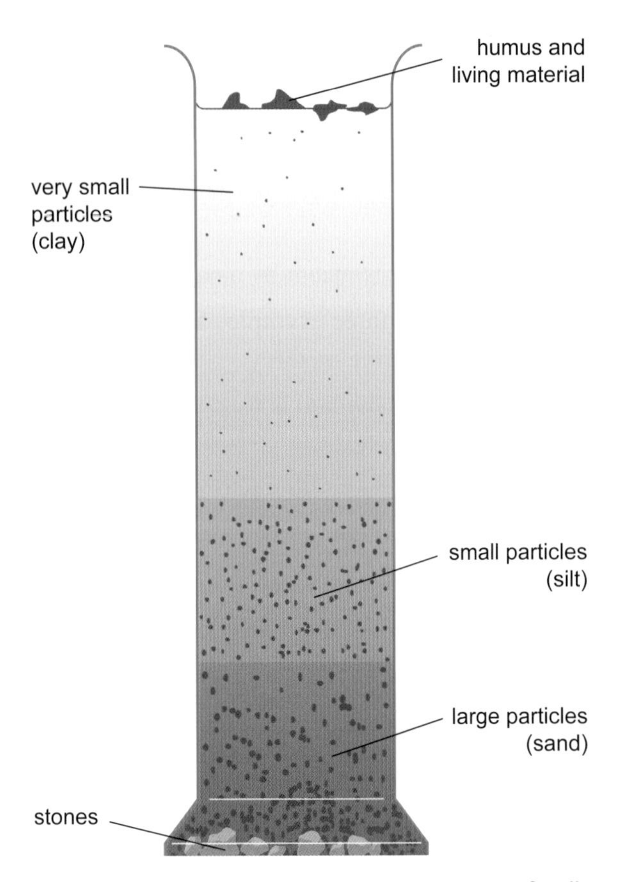

The contents of soil.

These soil crumbs contain tiny bits of broken rock and plant matter.

Gardeners dig compost and manure into soil. Compost is rotted plant material.

Earthworms make burrows that help get air into the soil.

Question 1 

Different soils for different plants

Spiking a lawn lets air into the soil, helping the grass to grow better.

Good gardeners put a lot of effort into looking after the soil so it is good for the plants they want to grow. Plants need a variety of things from the soil. These include:

- water;
- support;
- minerals;
- oxygen.

Most plants grow best in a soil that has a pH of between 6 and 8. Within that range some plants grow better in **alkaline soil** and some grow better in **acidic soil**. A soil with a pH of 8 is an alkaline soil. A soil with a pH of 6 is an acidic soil. The table shows some plants suited to different types of soil.

Cowslips grow well in alkaline soil.

Plants that grow well in acidic soil	Plants that grow well in alkaline soil
crocus	cowslip
rhododendron	lilac
camellia	flowering cherry
Chinese witch hazel	wallflower
good-luck plant	iris

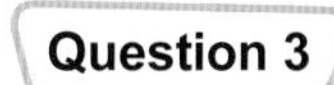

Question 3 4

Rhododendrons grow well in acidic soil.

Improving the soil

Gardeners and farmers add materials to the soil to improve it.

Compost is rotted plant material, usually from vegetable peelings, grass cuttings and dead leaves. This is added to the soil to increase the humus content, to help it hold water and stop it drying out. Digging compost in also helps to break up the soil and let air in. It also brings up minerals that have been washed down into the soil.

Fertilisers are added to increase the soil's content of minerals that plants need, such as nitrogen and potassium.

Soils are often too acidic for plants to grow well so people add crushed limestone or lime to the soil to neutralise the acid and raise the pH.

Crushed limestone can be spread on fields to make the soil less acidic.

Question 5 6

9G.2 Acid rain

You should already know | Outcomes | Keywords

Acid rain

One reason the soil can get too acidic for plants is that the rain that falls is very slightly acidic.

The **pollution** we produce from cars, power stations and burning fossil fuels contributes to the problem of **acid rain** but it is not the only cause. Living things breathing out carbon dioxide gas also contribute to the effect.

Acid rain can kill trees and the fish in lakes.

Ascension Island is in the middle of the Atlantic Ocean. It is a long way from anywhere. It is about as free from pollution as you can get! The drawings show what would happen if you collected some rainwater on Ascension Island and tested its pH with universal indicator.

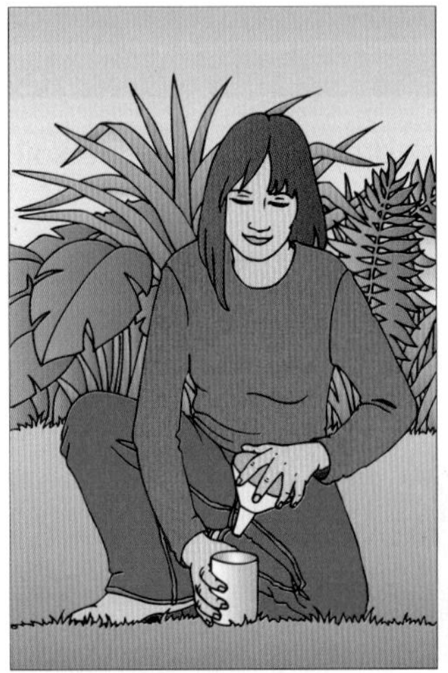

The pH of unpolluted rainwater is about 5.5. This shows that unpolluted rainwater is slightly acidic.

If you test the rain in a town in Britain you can get a pH reading of 4. This is evidence for the idea that the pollution in Britain has had a significant effect in making the rain more acidic.

Question 1 2

What is the effect of acid rain?

The photograph on the previous page shows the affect of acid rain on trees. Acid rain also affects rocks that contain calcium carbonate. This means it affects building materials made from thosc rocks. Thc tablc below lists some rocks and how they are affected by acid rain. The photograph shows the effect on a limestone carving on a building over a period of many years.

Example of rock	What happens to the rock in acid
granite	little or no change
sandstone with a silica cement	little or no change
sandstone with a carbonate cement	falls apart as the cement reacts with the acid
limestone	fizzes and disappears
chalk	fizzes and disappears

Acid rain affects metals

Even though acid rain is very dilute, over the years it will corrode metals like iron or zinc and alloys like steel. It does not affect lead. Lead is used in the building trade to stop water getting into sections of roof, particularly where sections of roof meet a wall. The sheets of lead used in this way are called lead flashing.

Acid rain is very dilute, but it has still corroded these metal railings faster than normal rain would.

How does acid rain affect animals?

Large animals like human beings are not directly affected very much by acid rain. Some of the things wc cat and drink, like orange juice and cola, are more acidic than acid rain. They have a pH round about 3.

Smaller animals like frogs and fish have a lower level of tolerance.

- Acid rain in rivers and ponds can stop eggs hatching.
- Acid rain can kill off the plants that animals eat.

At first the smaller, acid-sensitive living things disappear. Then larger living things that feed on them go too as the food supply diminishes. Gradually the number of living things reduces. This is called **progressive depletion**.

There is nothing alive in this river.

Check your progress

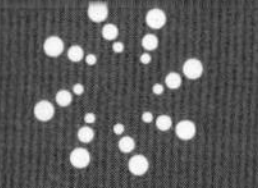

9G.3 Can we reduce the amount of acid in the rain?

You should already know | Outcomes | Keywords

What causes acid rain?

The photograph and the drawing show some of the things that cause **acid rain**.

Human beings contribute to it just by breathing out! Humans breathe out about 4% carbon dioxide gas. Some of this dissolves in rainwater in the atmosphere, producing a weak acid. All living things give off carbon dioxide so all living things make some contribution to acid rain.

Other natural events like the occasional volcano add to the problem. Volcanoes produce **sulfur dioxide** gas when they erupt which dissolves in the rainwater and makes it acidic.

Big contributions to acid rain come from cars, lorries and aircraft. Power stations that use fossil fuels also produce **carbon dioxide** and sulfur dioxide. These gases contribute to acid rain.

The idea of reducing your carbon footprint is about reducing the amount of carbon dioxide that your actions help produce. Every time you use electricity that comes from a power station using a fossil fuel, and every time you use a car or fly in a plane, you contribute to the problem.

Question 1  2

How do gases from fuels and cars make acid rain?

Fossil fuels contain small levels of sulfur as well as a lot of carbon. When the fuel is burned, carbon dioxide and sulfur dioxide are produced in the reaction. Sulfur dioxide is a gas. It passes into the air along with all the other materials and gases that go up the chimney.

When sulfur dioxide mixes with the oxygen and rainwater droplets in the air, it produces acid rain.

A similar process happens with the **nitrogen oxides** produced by car exhausts.

water droplets
sulfur dioxide
oxygen
sulfur dioxide gas is released into the air
droplets of sulfuric acid
burning fuels containing sulfur
the rain that falls is acid rain

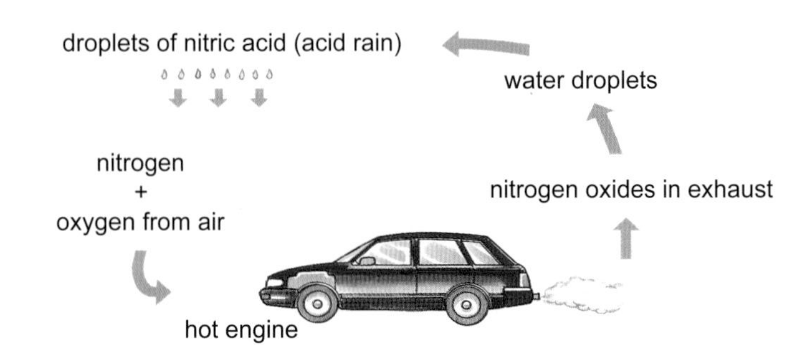

Question 3

Can we reduce the amount of acid rain?

There are things human beings could do to reduce the level of acid rain. Unfortunately, some these things tend to be inconvenient or represent an increase in costs. We could:

- use more energy from renewable sources and burn less fossil fuel;
- remove sulfur from fossil fuels before we burn them;
- remove the gases that cause acid rain from chimney smoke and exhaust fumes.

The photograph shows catalytic converters used to remove nitrogen oxides from car exhausts.

The catalytic converter is in the car's exhaust system. Catalysts on special surfaces inside the converter speed up the reactions that make the exhaust gases less polluting.

The substance inside the converter makes a reaction happen between the different gases in the exhaust fumes. The word equation is:

nitrogen oxides + carbon monoxide → nitrogen + carbon dioxide

The amount of carbon dioxide you produce is less of a problem than the nitrogen oxides you would produce without the catalytic converter.

The substance that makes this happen is called a catalyst. A catalyst is a substance that helps a reaction happen but does not take part in it.

One problem is that catalytic converters are expensive because they contain small amounts of expensive metal like platinum.

Question 4 5

How about the sulfur dioxide?

Catalytic converters do not remove sulfur dioxide. To reduce the sulfur dioxide levels in exhaust fumes, petrol with reduced levels of sulfur has been produced. This is often called 'cleaner fuel' at the petrol station pump.

Question 6

This pump contains ultra-low-sulfur unleaded petrol, so the exhaust gases from cars that use it will contain very little sulfur dioxide.

9G.4 Global warming

You should already know | Outcomes | Keywords

What evidence have we got for global warming?

Scientists have been collecting data for over a hundred years on things like air temperature, wind and rainfall.

Apart from these historic records of the weather, scientists also get evidence of the Earth's climate in the past from:

- tree growth rings;
- air trapped in the Antarctic ice;
- ocean sediments;
- ocean corals.

The graph below shows the global temperature in the recent past.

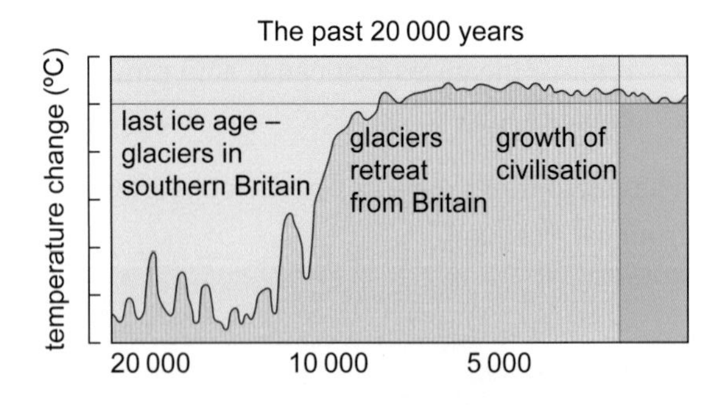

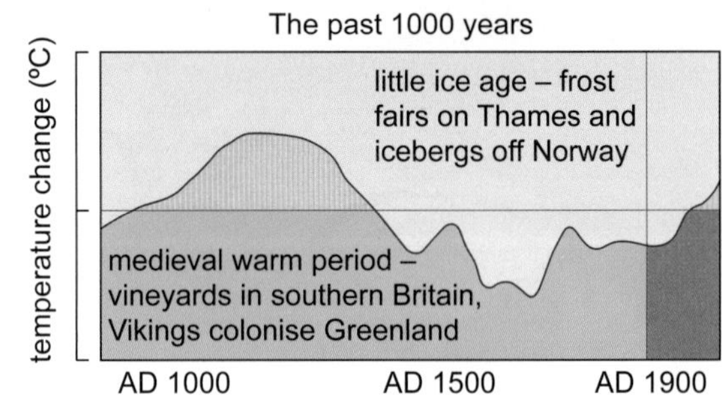

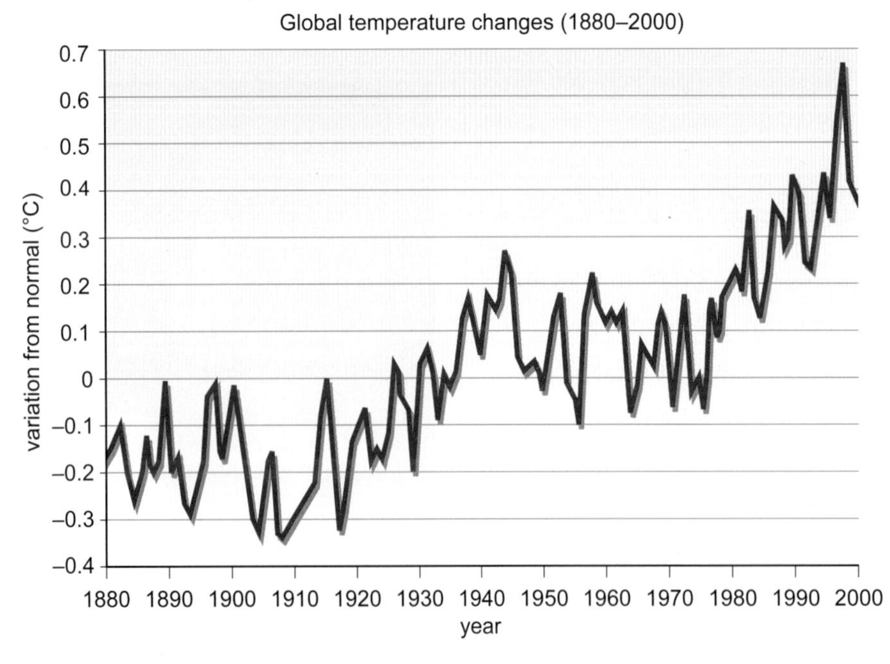

- The red lines show the average temperature in the second half of the 20th century.
- The scales on the two graphs are different.
- During the 'little ice age', it was so cold in the winter that the River Thames froze and people could take part in activities on the ice.

The graph shows that a temperature rise of about 0.5 °C happened between 1980 and 2000. This does not sound a lot but if it carries on then by 2100 the surface temperature of the Earth could easily rise by 5 °C, which would have major consequences for living things. We call this **global warming**.

Question 1 

How does the Earth warm up?

The Sun's rays heat the Earth all the time. The Earth does not get hotter because it radiates the heat back into space.

Some of the gases in the air reduce this heat loss. The gases trap the heat and stop it escaping into space, a bit like the glass in a greenhouse traps the heat.

We call the effect of gas in the Earth's atmosphere trapping heat, the **greenhouse effect**.

There are two important points about the greenhouse effect.

- It is a natural effect and without it the Earth would be a colder planet.
- One of the main gases that causes the greenhouse effect is carbon dioxide but other gases like methane make a contribution too.

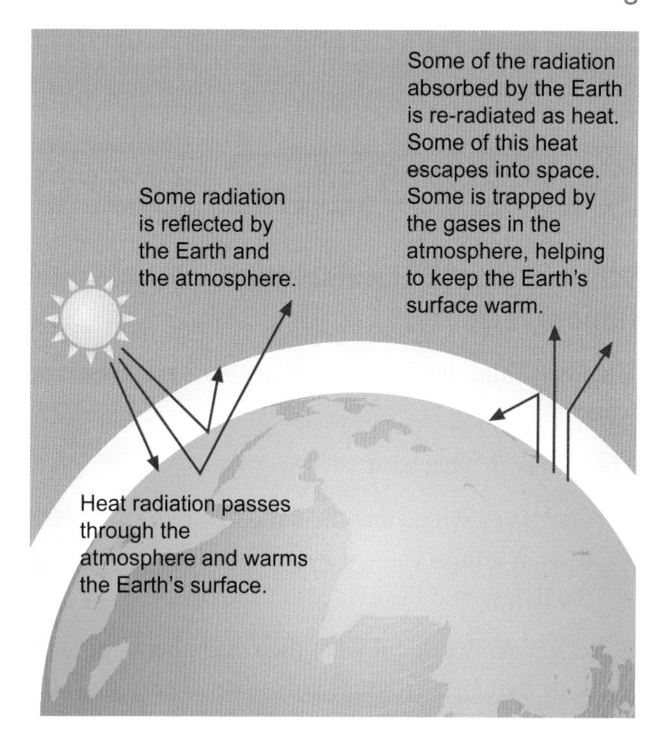

Before humans started to pollute the air by burning fossil fuels and developing industries, the level of carbon dioxide was thought to be about 0.03%. This is just right for a greenhouse effect that keeps the planet at a temperature suitable for humans. In recent times humans have increased the levels of carbon dioxide and other gases we call **greenhouse gases**. This has increased the greenhouse effect and the Earth's atmosphere is warming up.

Several areas of Britain were severely affected by flooding after storms in June 2007.

Question 3 4

What might happen?

If the world continues to warm up, there could be several effects.

- Weather patterns might change. This is often referred to as **climate change**. Winters will be less cold. Countries like the UK are experiencing greater extremes of weather, like floods and high winds, which have been less frequent in the past.
- Scientists believe that ocean levels could rise by 50 cm in the next hundred years. Cities like Alexandria and Port Said on the Nile Delta would be under threat if this happened.
- Patterns of wildlife will change. Sea life which is only found in warm waters to the south of Britain will migrate north and be found off the north of Scotland. Animals like polar bears that live in the Arctic will find the areas of ice a lot smaller than they are at present.

Global warming will damage the habitat of animals like the polar bear.

Review your work

Summary ➡

Question 5

Is global warming real?

You should already know

Outcomes

Keywords

Why do most scientists think global warming is happening?

Scientists think the Earth is getting warmer. This idea is called global warming.

They think this because:

- there is evidence that the temperature of the Earth is rising;
- there is a theory about why this is happening;
- the theory and the evidence fit together.

Scientists can work out what the atmosphere has been like in the past by studying evidence from ice cores. The theory that supports the idea that global warming is taking place is based on the fact that the changes in temperature for different years are similar to the changes in the levels of carbon dioxide trapped in the ice.
Look at the red and blue lines on the graph.

Scientists collect ice cores by driving a hollow tube up to 3 km into the ice. We can use the ice to find out about the climate and the atmosphere up to 750 000 years ago.

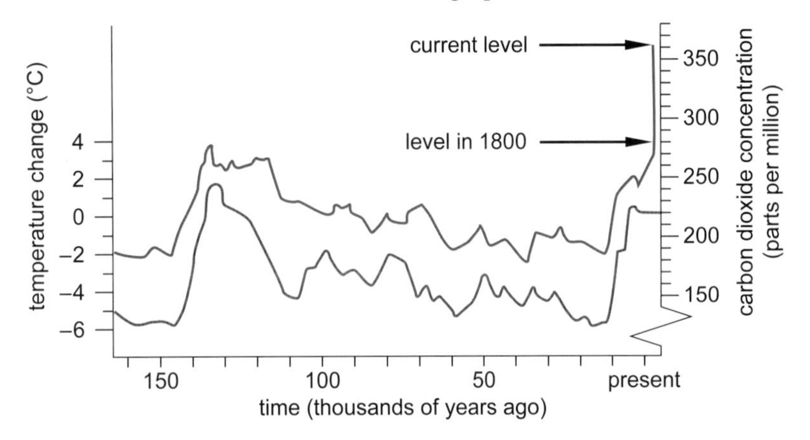

Key
— CO_2 concentration
— temperature change

Evidence from the ice cores tells us that there is a link between the temperature on the Earth and the concentration of carbon dioxide in the atmosphere.

Just because the graphs for carbon dioxide and temperature change follow the same pattern does not mean there is a link. Both graphs could be caused by something else. Because of this, scientists who believe that carbon dioxide is causing the temperature change need to be able to prove how that could happen. They need to be able to rule out **alternative explanations** for the temperature change.

One possible cause for the recent rise in carbon dioxide in the atmosphere is the way humans have been using fossil fuels. The pattern of use of fossil fuels fits in very well with the increase in carbon dioxide and the increase in temperature. This makes people think that fossil fuel use is a key factor in global warming.

Question 1 2

How good is the evidence?

Scientists say that the temperatures shown for the last 130 years are very **reliable** evidence. This is because:

- the temperatures have been directly measured with thermometers;
- the results are the averages of many temperature readings to eliminate errors.

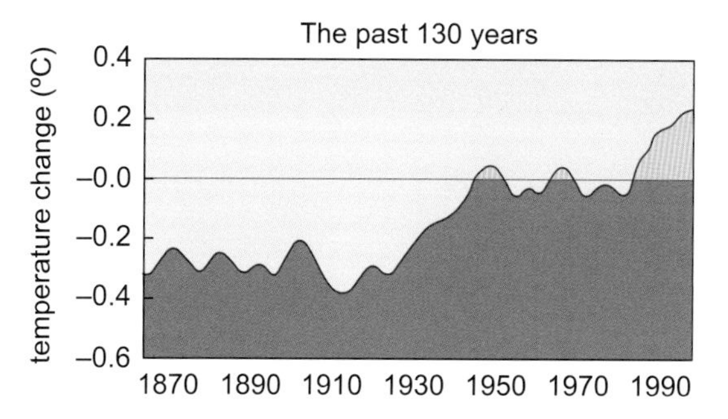

The red lines show the average temperature in the second half of the 20th century.

Earlier evidence might not be so reliable because it depends on **indirect observations** and theories about how things behaved in the past.

It seems clear that there has been a rise in temperature during the last hundred years. The question is, 'Is that temperature going to continue up, or is it just going to reach a peak and come back down like it did in the medieval period several hundred years ago. To answer this question we really need to be sure what the cause of the temperature change is.

Question 3

What is the theory that fits in with the evidence?

Energy reaches the surface of the Earth from the Sun as infra-red radiation.

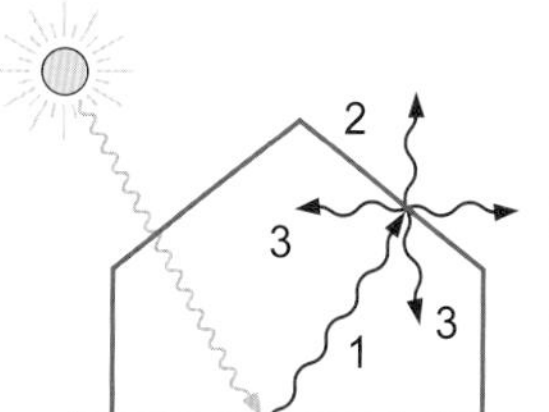

1 The soil becomes warm and radiates long wavelength infra-red rays.
2 The glass absorbs the infra-red radiation.
3 Some of the energy is re-radiated into the greenhouse which makes it warmer.

Everything gives off infra-red radiation. High temperature objects give off more than low temperature objects.

High energy radiation has a short wavelength. Low energy radiation has a long wavelength. The different types of infra-red rays can be detected with a photograph. The photograph of the horse shows the infra-red radiation coming from the horse. The hotter parts of the horse give off shorter wavelength waves. These show up as brighter parts on the image. You can see that the hooves are much cooler than the head.

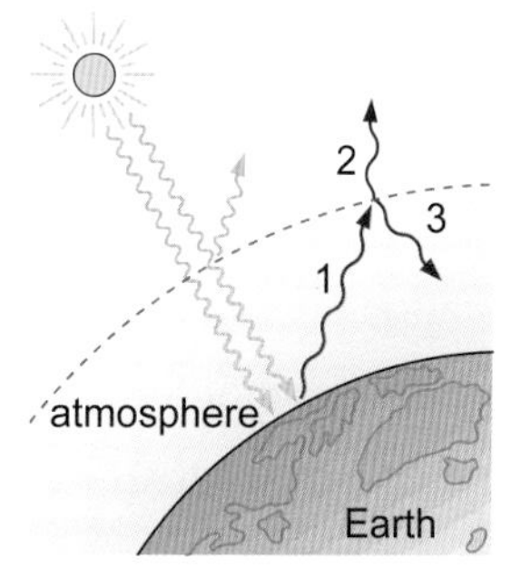

1 The Earth becomes warm and radiates long wavelength infra-red rays.
2 Greenhouse gases absorb some of the infra-red radiation. They re-radiate some of it back towards the Earth.
3 The atmosphere gets warmer.

The idea of infra-red waves can be used to make a prediction using the global warming theory because:

- The short wavelength waves from the Sun pass through the atmosphere and warm up the Earth.
- The warm Earth gives off long wavelength infra-red waves. Some of these are radiated back to Earth by the carbon dioxide in the atmosphere. This means that not all of the heat from the Earth escapes into space. The Earth gets warmer.
- The increase in carbon dioxide concentration in the Earth's atmosphere should therefore produce an increase in the Earth's temperature. This is a prediction that can be tested.

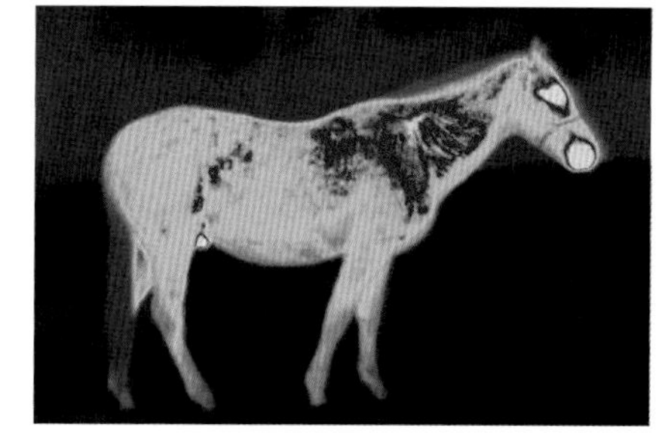

An infra-red camera shows that all objects give out infra-red rays. Hotter areas appear brighter on the image.

9G Questions

9G.1

1 a What is humus and what does it do to the soil?

b Give an example of something that helps form humus.

2 How do worms help with the structure of the soil?

3 State three substances that plants get from the soil.

4 a What is the difference between alkaline soil and acidic soil, and what effect does this have on plants?

b What is the range of pH for which most plants grow well?

5 What is compost and why do gardeners add it to the soil?

6 Give one method that is used on farms to change the pH of the soil from being too acidic.

9G.2

1 a How does the pH of rainwater compare between Ascension Island and the UK?

b What does the difference suggest?

2 What can cause rainwater to become more acidic?

3 a Give one example of a rock that is not affected by acid rain and two examples of rocks that are affected.

b What implication does this have for building materials?

4 a Which metal used in building is not affected by acid rain?

b Give one example of where this metal is used.

5 a Describe what is meant by progressive depletion.

b What could be done to prevent it happening?

9G.3

1 Describe <u>two</u> natural events that contribute to acid rain and describe how they do this.

2 What does 'reducing your carbon footprint' mean?

3 a Why does burning a fossil fuel produce sulfur dioxide as well as carbon dioxide?

b Why does this contribute to acid rain?

4 a What can humans do to reduce the levels of acid rain?

b Why might humans not want to do these things?

5 a What does a catalytic converter do to exhaust gases?

b Why is this useful?

c Give <u>one</u> disadvantage of using a catalytic converter.

6 What can be done to reduce the level of sulfur dioxide produced by car fuel?

9G.4

1 What sources of evidence do scientists use to work out what the Earth's climate was like in the past?

2 a Describe how the temperature on the Earth has changed in the recent past.

b What does this pattern of change mean if it continues and what is the term used to refer to it?

3 What is the greenhouse effect and what is its impact on the climate?

4 Name <u>two</u> greenhouse gases.

5 Give <u>four</u> possible consequences of continued global warming.

9G.HSW

1 Why do most scientists think the idea of global warming is correct?

2 What do scientists who believe that there is a link between carbon dioxide and global warming need to do to support their case?

3 Why are the records of temperature from the last 100 years more reliable than earlier evidence?

4 How can the idea of infra-red rays be used to explain global warming?

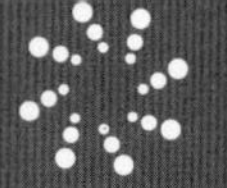

Scientific enquiry

9H.1 What happens when fuels burn?

You should already know | Outcomes | Keywords

Combustion

Burning is a **combustion** reaction. Combustion is:

- a rapid reaction;
- a reaction with oxygen from the air;
- a reaction which makes new substances called oxides.

Oxides are substances that contain oxygen and another element. Oxygen is a colourless gas so it can be hard to see what happens in a combustion reaction. **Fuels** are substances that produce useful levels of heat in a combustion reaction.

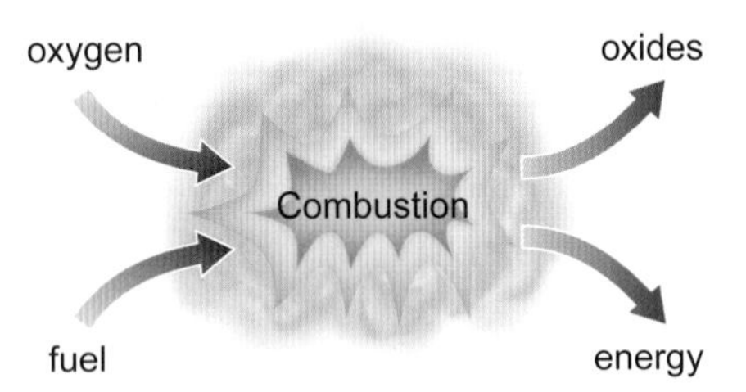

Fuels are chemicals that react with oxygen to give useful energy. Energy is always released when a fuel burns.

What happens when a fuel burns?

Barbecue charcoal is made from carbon. When it burns, it reacts with the oxygen in the air to make carbon dioxide. The word equation is:

carbon + oxygen → carbon dioxide

In the reaction each carbon atom joins with two oxygen atoms. The symbol equation is written:

$C + O_2 \rightarrow CO_2$

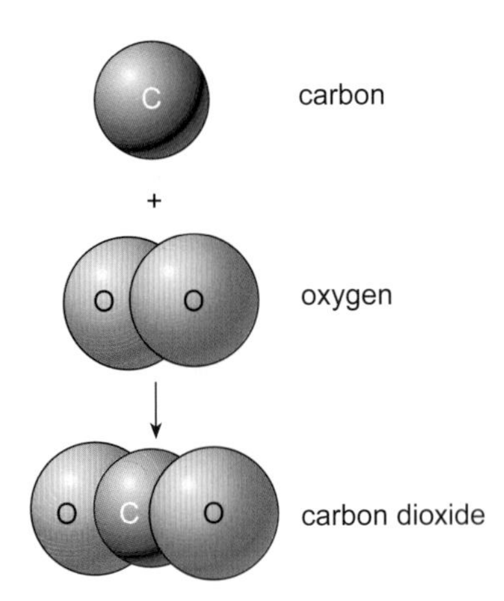

Carbon combining with oxygen.

The drawing below shows what happens when methane burns. Methane is the gas used in Bunsen burners and in domestic gas fires.

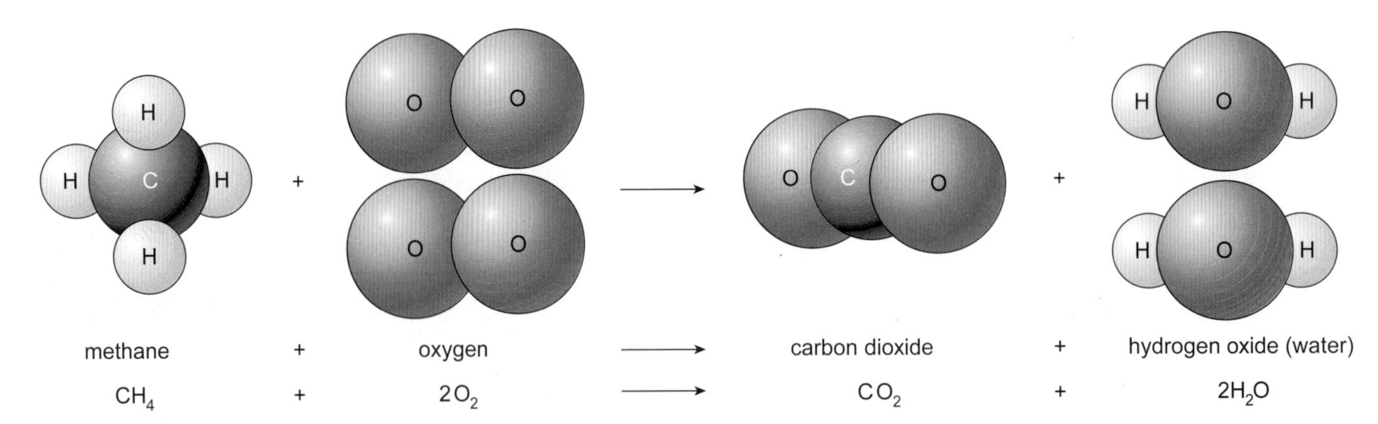

When methane burns, it joins up with oxygen in the air to form the oxides carbon dioxide and water.

Question 1 2

Not completely burnt

Sometimes there is not enough oxygen for a fuel to burn as well as it can. When this happens it is called **incomplete combustion**.

You get complete combustion of the gas in a Bunsen burner when the air hole is fully open and the flame roars. In this case the reaction is:

methane + oxygen → carbon dioxide + water

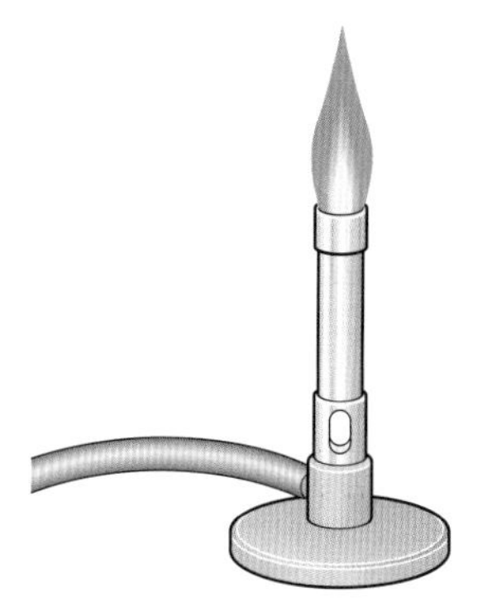

Water and carbon dioxide are formed when the air hole is open.

You get incomplete combustion when the air hole is not fully open. This is because there is less oxygen available to take part in the reaction. You can see a difference in the flame – it becomes almost invisible and does not make much noise. The reaction is:

methane + oxygen → carbon monoxide + water

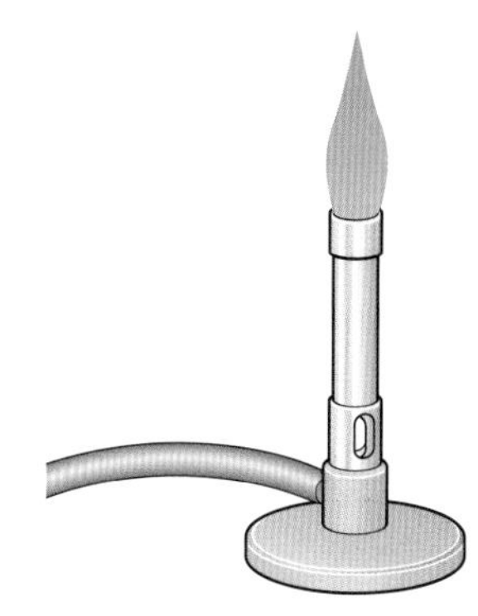

With the air hole half open, water and carbon monoxide are formed.

Carbon monoxide has the formula CO. It is a colourless gas just like carbon dioxide – except that carbon monoxide is poisonous!

The gas has no smell so it can be very dangerous. If a gas fire does not have the correct air supply, it will produce carbon monoxide which can kill. People use carbon monoxide sensors and have the fires serviced regularly to prevent accidents.

Closing the air hole

When you close the air hole on a Bunsen burner you get a yellow flame. This happens because the combustion reaction produces carbon rather than carbon monoxide when the oxygen is very limited. The reaction is:

methane + oxygen → carbon + water

You can detect carbon in a yellow Bunsen flame with a white tile. The particles of soot in the flame get very hot. This makes a yellow flame.

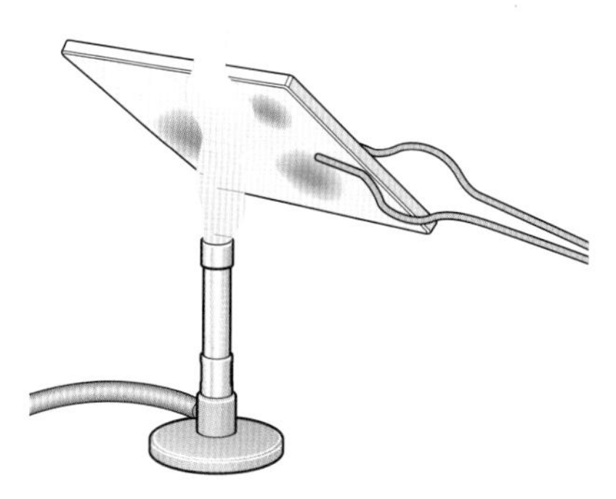

Carbon in the form of soot is made when the air hole is closed. Look at the soot on the white tile.

Matches, fireworks and explosives

These things are made from chemicals that contain the oxygen they need for combustion. The chemicals containing oxygen are called **oxidising agents** because they give oxygen to other chemicals.

All these things contain an oxidising agent to provide the oxygen for the reaction.

Question 5

9H.2 Chemical reactions for energy

You should already know | Outcomes | Keywords

Making electricity

Chemical reactions make new substances. They sometimes release energy. Often this is thermal energy that heats up the surroundings. Some chemical reactions can be used to release energy that will make an electric current flow in a circuit. We use these chemical reactions in cells and batteries.

There are many different types of **cell**. They are often named after substances that are in them, like nickel–cadmium, lead–acid or lithium.

The substances zinc, copper, lead, manganese, mercury and silver are commonly used to make cells.

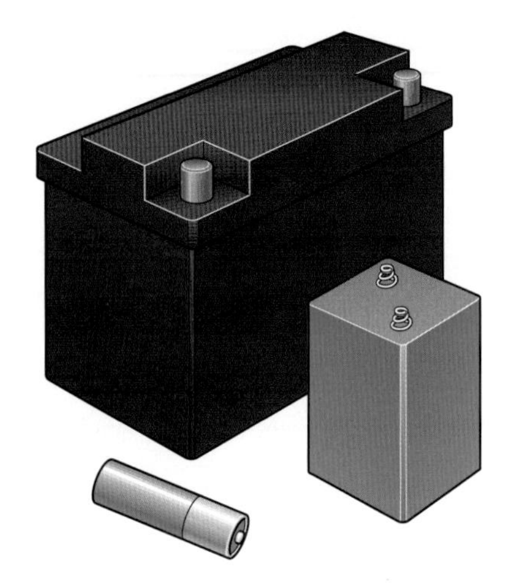

Cells and batteries.

Question 1 2

Different voltages

If you put a rod of magnesium and a rod of zinc into some sulfuric acid you can produce electricity.

If you connect a voltmeter to the rods you can see that the chemical reaction between magnesium, sulfuric acid and zinc produces a voltage of just under one volt.

The rods of metal are called **electrodes**. The rod connected to the red terminal of the voltmeter is called the positive electrode. The other rod is called the negative electrode.

Different pairs of metals dipped into the acid give different voltages. The table gives some examples

Negative electrode	Positive electrode	Voltmeter reading
magnesium	copper	1.91 V
magnesium	iron	1.42 V
magnesium	zinc	0.95 V

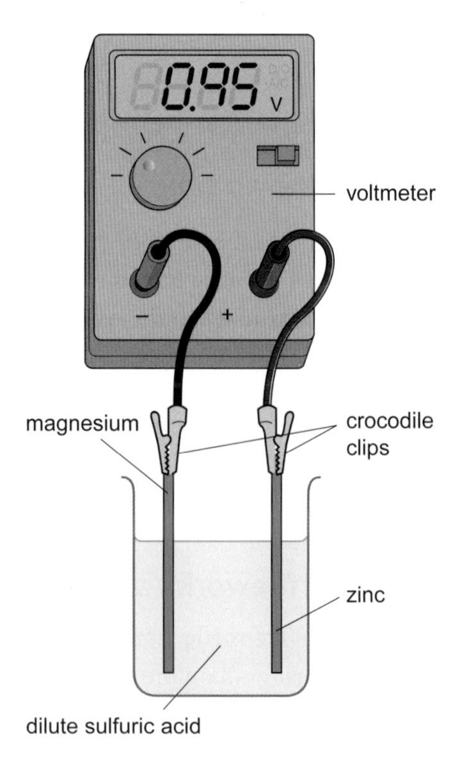

Different pairs of metals dipped in acid produce different voltages.

Question 3  4

Producing heat

When a chemical reaction happens, new substances are formed. Sometimes thermal energy is transferred to the surroundings. The everyday expression for this is to say that heat is produced. We use this effect when a fuel is burned to heat something up.

When a fuel burns it combines with oxygen from the air. This is an example of a combustion reaction.

We burn substances like coal because they produce heat. We call them fuels.

Question 5

There are many other reactions in which heat is produced that are not combustion reactions. Examples include:

- adding water to calcium oxide (lime);
- pouring sodium hydroxide into hydrochloric acid;
- mixing water with plaster and leaving it to set.

The diagram shows that sodium hydroxide and hydrochloric acid produce heat when they are mixed. The reaction between sodium hydroxide and hydrochloric acid is a neutralisation reaction.

Sometimes you see the fact the reaction transfers energy to the surroundings added to the word equation, like this:

hydrochloric acid + sodium hydroxide → sodium chloride + water + heat

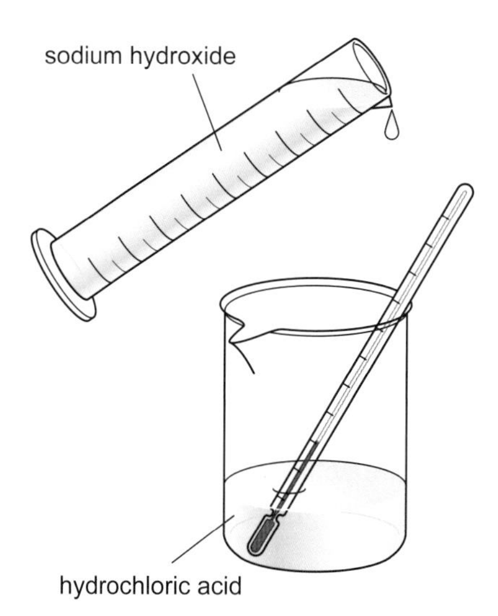

Adding the right amount of alkali to an acid can neutralise it. The reading on the thermometer increases during the reaction. Neutralisation is an exothermic reaction.

A reaction that transfers thermal energy to the surroundings is called an **exothermic reaction**.

Exothermic reactions are sometimes used to warm things up.

You can buy hand warmers that produce heat when a chemical reaction inside them is activated.

It is also possible to buy self-heating food. The drawing shows tins of coffee. When the button on the base is pushed, a chemical reaction between water and lime is started. This reaction is exothermic.
The word equation is:

calcium oxide + water → calcium hydroxide + heat

The energy transferred is enough to heat the coffee in the tins to 60 °C.

Question 6 **7**

Check your progress

You should already know | Outcomes | Keywords

Reactions make new materials

There are many different branches of chemistry. They all have their own name. Chemists working in each area study how substances in that area react and try to make new substances that are useful.

Examples include:

- **biochemistry** – substances in living things;
- **petrochemistry** – substances made from oil and gas, like fuels and plastics;
- **polymer chemistry** – development and manufacture of plastics and fibres;
- **pharmaceutical chemistry** – development and manufacture of drugs and medicine;
- **agrochemistry** – making fertilisers, weed killers and insecticides.

Once chemists have worked out how to make a useful product, it may well be manufactured on a large scale at an industrial plant like the one shown in the photograph.

This site produces many different types of chemicals.

Question 1 | 2

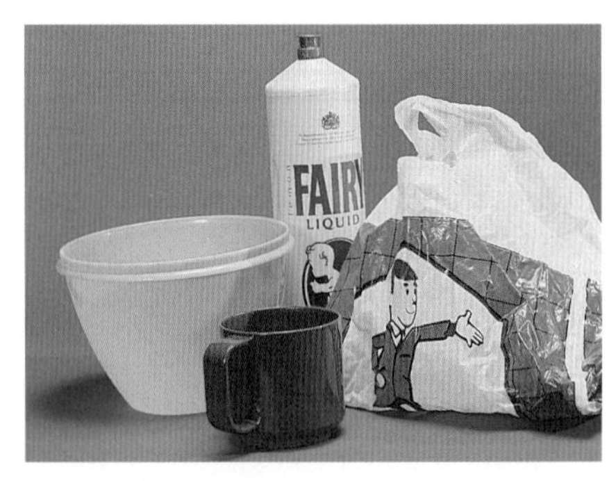

Many of the things we use are made by chemists working with polymers.

Oil from sunflowers is reacted with hydrogen to make margarine.

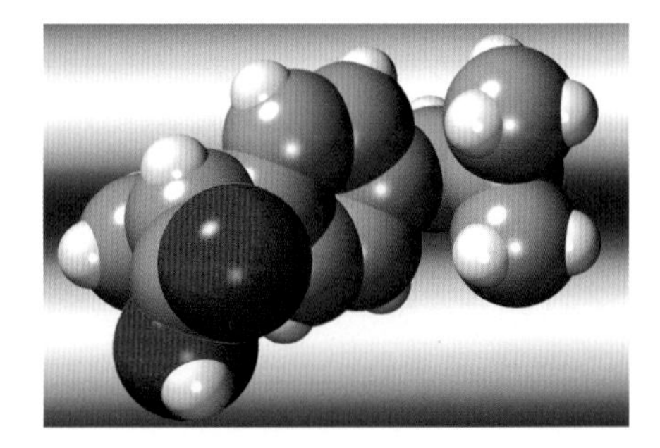

A model of a molecule of the painkiller ibuprofen.

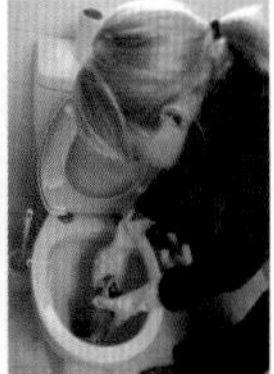

The site's products include chemicals for use in cosmetics, agriculture and cleaning products.

Using plants

Plants are excellent sources of raw materials. Plants have been used as medicine for thousands of years because of the medicinal properties of substances they contain. People have not always known what the plant contains but they have known of its beneficial effects.

- Willow bark and leaves were used 2500 years ago in Greece as a painkiller.
- The bark of the cinchona tree can be used to cure malaria.
- The leaves of the coca plant can be used to make the painkiller, cocaine.

We now get anaesthetics and cocaine from the coca plant.

Question 3

How we got aspirin

In the early 19th century, scientists discovered that the chemical in the willow bark that made it work as a painkiller was the substance salicin. This is known as the active ingredient.

Salicin had side effects if people swallowed it. It made some people feel sick and gave them bad indigestion. In 1899 some chemists made a new substance by reacting salicin with other chemicals. This stopped pain but did not have as many side effects. The new substance was named 'aspirin'.

The new substance was tested by volunteers in what we call a drug trial. Often drugs are tested on animals before human volunteers try them out.

Making new substances from others is called **synthesis**. The new material is called a **synthetic material**.

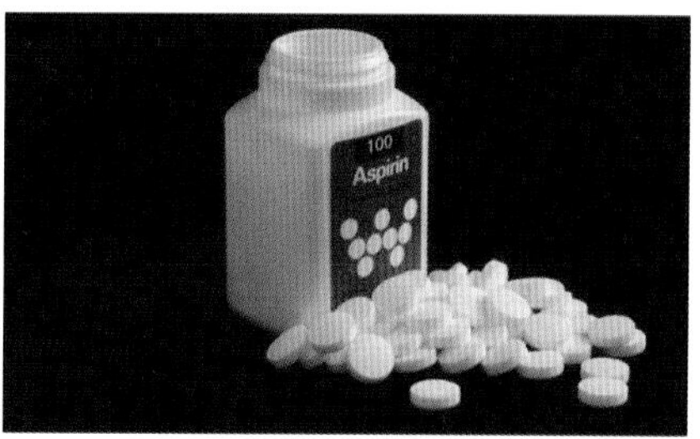

The painkiller we call aspirin was originally extracted from the bark of willow trees. It is now made from synthetic chemicals.

Question 4 5

Looking after the planet

The standard of living we have today depends upon many of the new substances made by scientists. Fertilisers and insecticides help produce enough food. Medicines help people to survive. Plastics and other synthetic materials are used to improve the quality of life in general by their use in clothing and household goods.

In improving the way we live, humans destroy large areas of plant and animal life. This means amongst other things that some species of plants and animals die out. This also means that a source of substances for new drugs is lost to us. The idea of making things better but <u>not</u> destroying the resources is called **sustainable development**.

The Amazon rainforest is a great source of new materials for drugs. But 1% of it is destroyed by humans every year.

Question 6 7 8

9H.4 Making and breaking bonds

You should already know | Outcomes | Keywords

About bonds

A molecule of oxygen is made from two oxygen atoms joined together. We write this as the formula O_2. We call the join between the atoms a **bond**. A bond is not an extra thing like the mortar between bricks. It is a word we use to mean that atoms are attached to each other in a similar way to a magnet and a piece of iron sticking together.

When a reaction happens, bonds are broken between some atoms and new bonds are made as the atoms rearrange into new substances.

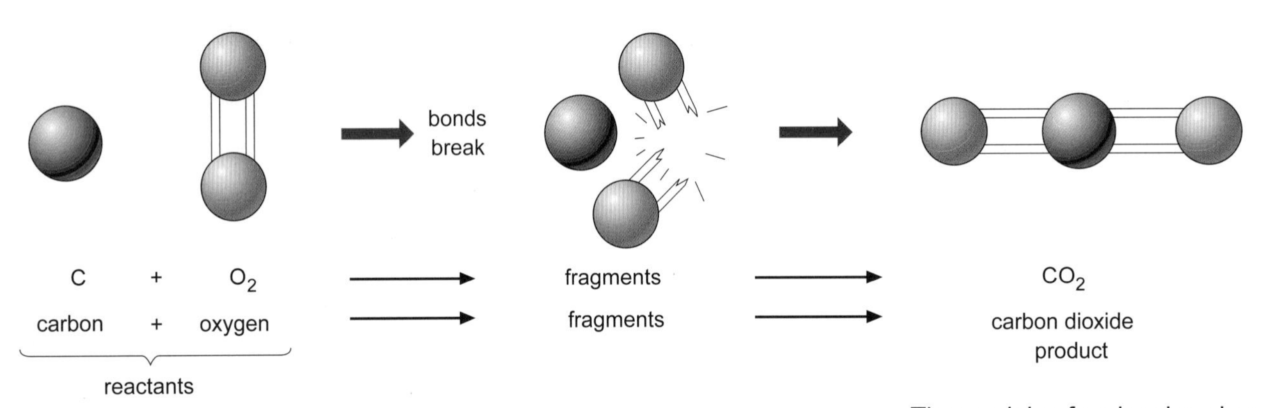

The models of molecules show what happens when barbecue coal burns. Bonds are broken. Then new bonds are formed to make carbon dioxide.

Question 1 | 2

Energy

You need to put in **energy** to break bonds. One way to do this that works for some reactions, is to warm up the reactants to get the bonds to break so the reaction happens.

When the bonds form, the energy that comes out of the reaction can be greater than the energy needed to get the bonds to break at the start of the reaction. If this happens, you get out more energy than you put in. This means the reaction gives out energy and there will be a rise in temperature. We say the reaction is **exothermic** because it gives out heat.

The reaction between carbon and oxygen is exothermic – it gives out heat.

Question 3 |  4 | 5

The number of particles

The total number of atoms at the end of a reaction is the same as the total number of atoms at the beginning. The atoms have just been rearranged.

The word and symbol equations for barbecue charcoal burning are:

carbon + oxygen → carbon dioxide
$C + O_2 \rightarrow CO_2$

At the start of the reaction there is one carbon atom and two oxygen atoms. At the end of the reaction the same atoms are still involved but they have joined up to make a carbon dioxide molecule.

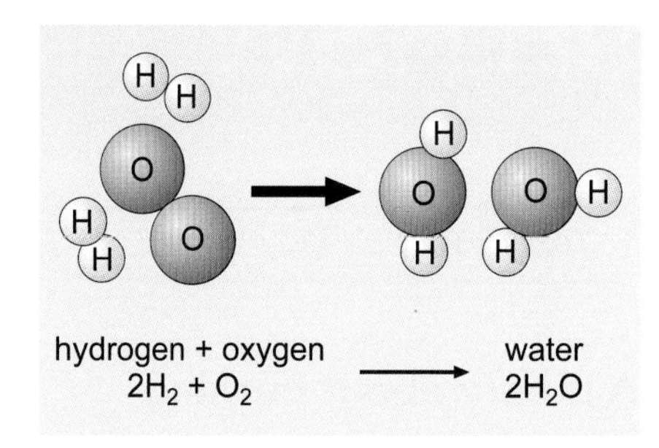

When two molecules of hydrogen react with a molecule of oxygen, two molecules of water are produced. The total numbers of atoms do not change.

Question 6

The mass stays the same

The mass of a substance depends upon how many atoms it is made from. The number of atoms at the end of a chemical reaction is the same as the number there were at the beginning. This means that the mass of the substances at the end of a reaction is the same as the mass of the substances at the start. No mass is lost in a reaction. This is an important law in chemistry and it is sometimes called the law of **conservation of mass**.

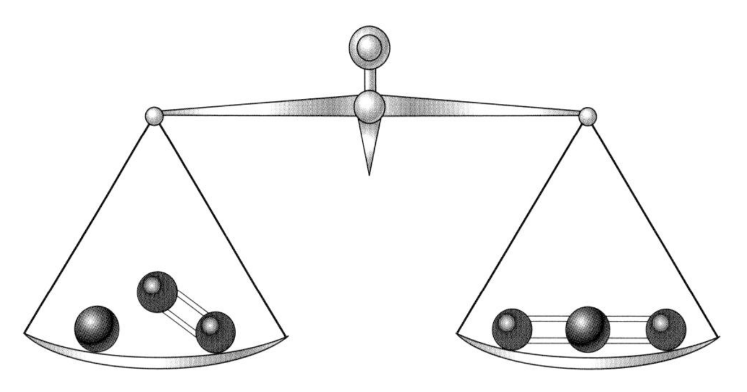

No mass is lost when reactants change into products.

Question 7

Don't forget about the escaping gas!

If you try the experiment in the diagram below without covering the candle, the mass shown on the balance will drop.

It seems that mass is lost in the experiment. In fact, the carbon dioxide gas produced in the reaction has escaped and its mass is not contributing to the reading on the balance. It is important to remember that, even though a gas is invisible, it still has a mass.

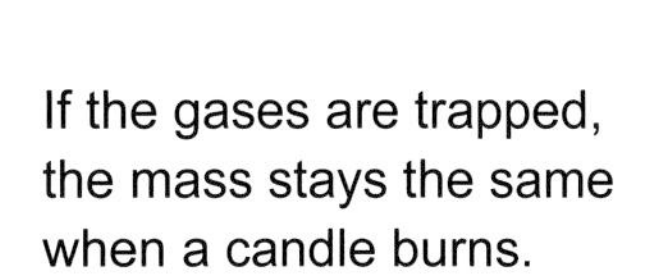

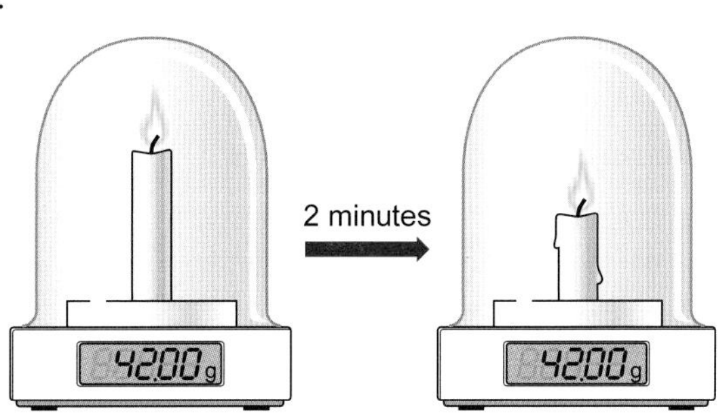

If the gases are trapped, the mass stays the same when a candle burns.

Question 8

Review your work

Summary ➡

9H.HSW A theory of combustion

You should already know

Outcomes

Keywords

A theory about burning

Burning wood seems to show a large loss of mass.

When people first tried to explain what happened when something burned, they got it very wrong. It seemed obvious that when something burned there was less at the end than there had been at the beginning.

You can see this if you set fire to a pile of wood. It eventually reduces to a pile of ash. Because of this, scientists at the time concluded that when a substance burned, something escaped.

This idea also fitted in with the observation of flames and smoke, which seemed to be leaving the burning substance.

The first **theory** was produced about 1690 from the ideas of two German chemists called Johann Becher and Georg Stahl.

The idea was that anything that could burn contained a substance called **phlogiston**. When something burned, it was the phlogiston that escaped from the substance in the form of flames and light.

Question 1

Another observation

If you put a lighted candle under a closed container, it goes out after a few minutes. The explanation for this in 1700 was that the air around the candle could only hold so much phlogiston. Once the air in the container was full of phlogiston, the candle would go out.

This fitted in with the observation that if you lift the cover and let fresh air in, you can light the candle again.

Pure air was called **dephlogisticated air** because scientists thought it did not have any phlogiston in it.

Question 2

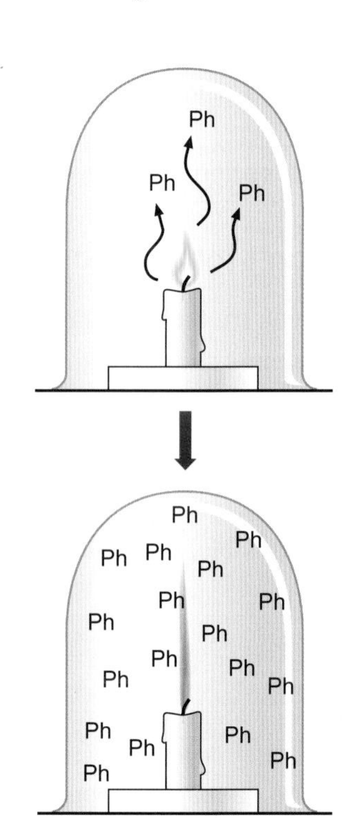

When the candle burns, people thought it gave out phlogiston. When it went out, they thought the air was full of phlogiston.

Oxygen

In Sweden, in 1772, a scientist called Carl Scheele made a gas which he called 'fire air'. We now call it **oxygen**. Two years later, another scientist called Joseph Priestley also discovered oxygen without knowing about Carl Scheele's discovery. Priestley did not know about Scheele's discovery because news travelled a lot more slowly in 1770 than it does today. There was no television, radio, telephone system or internet. Priestley called oxygen 'dephlogisticated air' because things burned very easily in it and he believed it was the purest form of air. It was as if it could soak up more phlogiston than any other gas or combination of gases.

Carl Scheele, 1742–1786.

The right idea about burning

The first to person to interpret the evidence about burning correctly was a French scientist called Antoine Lavoisier.

At first, Lavoisier believed the phlogiston theory was true but when he started weighing things carefully before and after burning, he noticed something interesting. He noticed that substances became slightly heavier when they were burned. This meant that phlogiston must have a negative mass! Lavoiser's observations followed from his use of an accurate balance that could weigh to 0.0005 g.

Antoine Lavoisier lived during the time of the French Revolution. It was a time when many old ideas and theories were challenged.

From his experiments, Lavoisier proposed the idea that mass is conserved in a reaction. He also worked out the burning is not phlogiston being released but a substance combining with oxygen. He published his ideas in 1777.

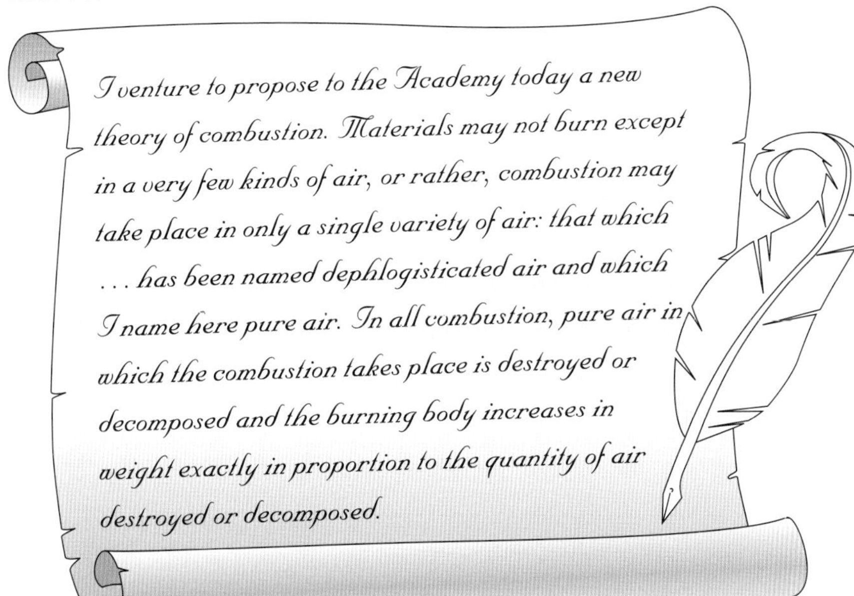

I venture to propose to the Academy today a new theory of combustion. Materials may not burn except in a very few kinds of air, or rather, combustion may take place in only a single variety of air: that which ... has been named dephlogisticated air and which I name here pure air. In all combustion, pure air in which the combustion takes place is destroyed or decomposed and the burning body increases in weight exactly in proportion to the quantity of air destroyed or decomposed.

This paper destroyed the theory of phlogiston. Soon no-one believed in phlogiston. All scientists now agreed that oxygen was used in combustion.

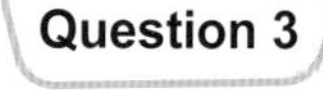

9H.1

1 Describe what is meant by a combustion reaction.

2 Describe the reaction that takes place when carbon burns in air.
Give the word and symbol equations for the reaction.

3 a How do the products of complete combustion differ from the products of incomplete combustion?

b Why does incomplete combustion happen?

4 a What is the symbol for carbon monoxide?

b What can be done to prevent people being poisoned by carbon monoxide gas?

5 a What causes a Bunsen burner to produce a yellow flame?

b How can you demonstrate that a substance is produced in the flame?

c What substance is it?

6 a What is an oxidising agent?

b Give three examples of situations where oxidising agents are used.

9H.2

1 a Name one thing that a chemical reaction always produces.

b What else do some chemical reactions produce that can be useful?

c Give some examples.

2 How are different types of cell sometimes named?

3 Describe a chemical reaction that can be used to produce electrical energy.

4 How can you change the reaction to produce a different voltage?

5 a Give one example of a reaction which transfers thermal energy to the surroundings.

b What is the name for a reaction that does this?

6 Give three examples of exothermic reactions.

7 Give one practical application of an exothermic reaction.

9H.3

1 Give the names of four different branches of chemistry and briefly describe the differences between these areas.

2 If chemists discover a useful product that is then needed in very large quantities, what might happen?

continued

3 Give two examples of medical problems that can be treated with plants or their extracts.

4 a Describe where we get aspirin from.

b What is the key difference between aspirin and the active ingredient found in willow bark?

5 What is meant by a synthetic material?

6 Suggest some properties of plastics make them useful for clothing and household goods.

7 What is the problem with humans destroying large areas of plant life and animal life to improve the quality of life for human beings?

8 What is meant by sustainable development?

9H.4

1 a What is a bond?

b Give one example of a situation where a bond occurs.

2 a What happens to the bonds in a reaction?

b Describe what happens to the bonds when carbon burns.

3 What do you have to supply in order to get bonds to break?

4 What is given out when bonds form?

5 a What is the name of the type of reaction in which the energy given out by the bonds forming is bigger than the energy you put in the first place?

b How do you know this has happened?

6 What is always true about the number of atoms at the start of a reaction and the number of atoms at the end?

7 What is the law of conservation of mass and why is it true?

8 a Why might an experiment involving the production of a gas seem to break the law of conservation of mass?

b Give one example of such an experiment.

9H.HSW

1 What was the first theory about burning?

2 a How does the phlogiston theory explain a candle going out when you cover it with a jar that shuts out the air?

b What is the correct explanation?

3 a What improvement in technology enabled Lavoisier to make an observation that helped him question the phlogiston theory?

b What else did the technology allow Lavoisier to discover?

4 What conclusion did Lavoisier reach which explained burning correctly?

9I.1 Using energy

You should already know | Outcomes | Keywords

Energy makes things happen

When things happen, **energy** changes from one type or form into another, or it moves from one place to another. When it changes form, we say it is transformed. We say it is transferred when it moves from place to place.

What happens	What happens to the energy
an electric drill turns	electrical energy is transformed to kinetic (movement) energy by the motor in the drill
a radio produces sound	electrical energy is transformed to sound energy by the loudspeaker in the radio
a light bulb lights up a room	electrical energy is transformed to heat and light by the wire in a bulb; the light energy transfers from the bulb into the room

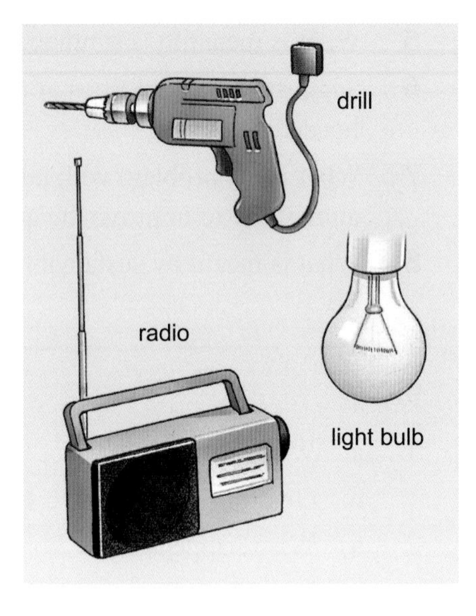

Question 1  

Storing energy

Some types of energy can be stored for later use.

Fuels are stores of energy. The energy in a fuel changes to thermal energy when it is burned. The type of energy stored in fuel is **chemical energy**. Wax, petrol, coal, oil and gas are all fuels – they are stores of chemical energy.

Coal, oil and gas are called fossil fuels because they come from the remains of plants and animals that lived millions of years ago. The energy in them originally came from the Sun. It was stored in the plants millions of years ago by **photosynthesis** when the plants were alive.

The energy in a cell or battery is stored as chemical energy. When a cell or battery is used, chemical energy changes to electrical energy.

Energy is stored as chemical energy in candle wax.

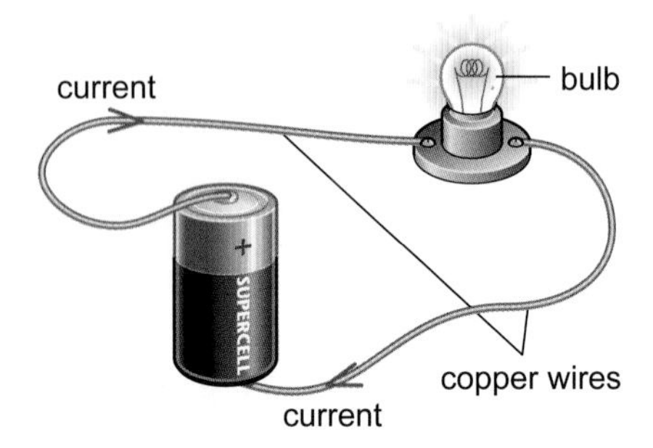

Chemical energy stored in the cell makes the current flow.

Question 4  

Potential energy

Potential energy means stored energy. The term is often used to refer to energy stored in an object when:

- its shape has been changed (e.g. a spring when it is wound up);
- it has been lifted up (e.g. a person at the top of a slide ready to go down it).

Elastic potential energy is energy stored when you change the shape of something, like stretching a spring or an elastic band.

The potential energy in something that has been lifted up is called **gravitational potential energy**.

The stretched springs store elastic potential energy.

Water stored behind dams has gravitational potential energy. This energy is transformed into electrical energy when the water is released from the dam, as shown in the diagram.

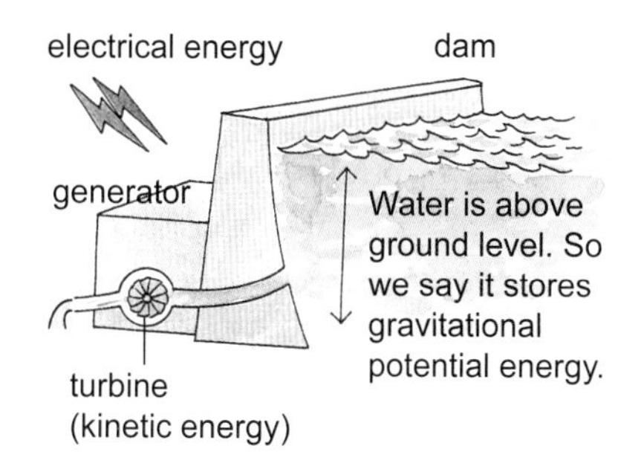

The change to kinetic energy

A common energy change is from potential energy to kinetic (moving) energy.

When an arrow is fired from a bow, the energy from the archer is first transferred to the bow by changing its shape. When the bow is released, the elastic potential energy stored in the bow is transferred to the kinetic energy of the arrow.

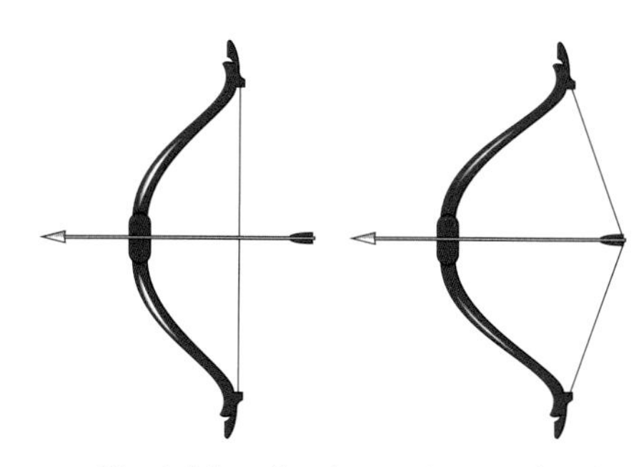

Stretching the bow stores elastic energy, which is transferred to the arrow.

The energy stored in the English longbows at the Battle of Agincourt in 1415 was enough to make arrows penetrate armour at a range of 100 metres. Scary! The longbow was the most feared weapon of its time.

Gravitational potential energy changes to kinetic energy when something falls.

Other example include:

- a cyclist freewheeling downhill from a high point to a lower point;
- the fall of the weight driving the pendulum in a pendulum clock.

When you wind up the clock, you lift up the weights. Energy stored by the weights is transferred to the clock as the weights slowly fall down.

Question 9

Check your progress

9I.2 Transferring energy

You should already know | **Outcomes** | **Keywords**

The current in a circuit

You have already met two types of circuit.

Series circuit	Circuit A is a series circuit. The current is driven round the circuit by the cell. The same current goes through each bulb in turn. As it flows through the cell, chemical energy from the cell changes to electrical energy in the current. As it flows through the bulbs, electrical energy from the current changes to heat and light energy in the bulb.
Parallel circuit	Circuit B is a parallel circuit. Each bulb has its own connections to the cell. The cell supplies electrical energy to separate branches in the circuit. Each branch carries electrical energy to a bulb where it changes into heat and light energy.

Circuit A.

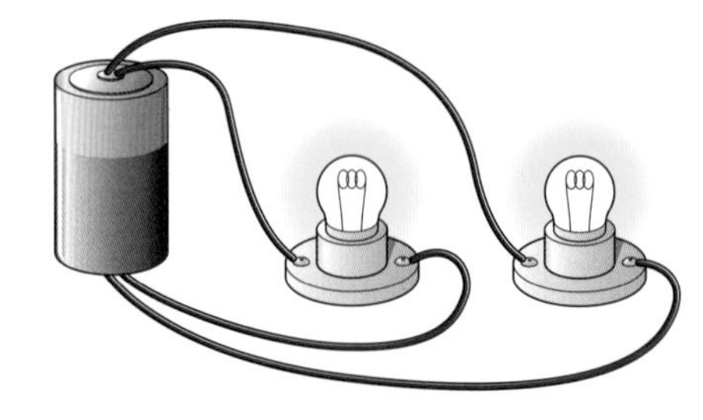

Circuit B.

In any circuit using bulbs and cells, all the energy given out by the bulbs comes originally from the chemical energy in the cells.

Question 1 **2**

Where does the cell's energy come from?

You can make a cell using two different metals, called the electrodes, and a solution that conducts electricity, called an electrolyte. The photograph shows how this can be done with coins and a lemon.

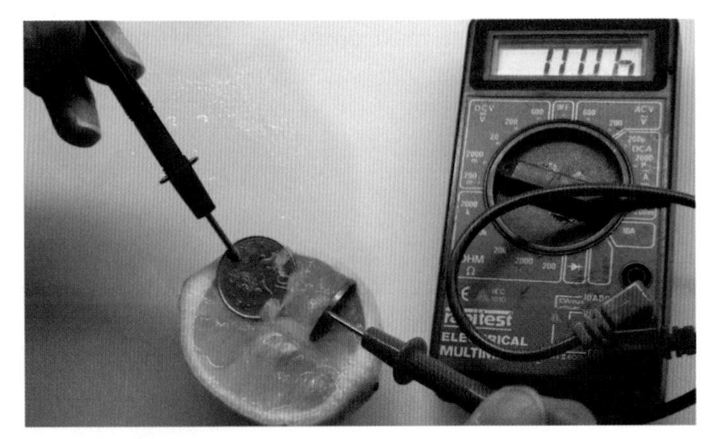

A cell's energy comes from the reaction between the substances in the cell. Chemical energy changes into electrical energy, which is carried by the current. The **voltage** tells us how much energy the cell can transfer to the current.

Question 3

What does the voltage tell us?

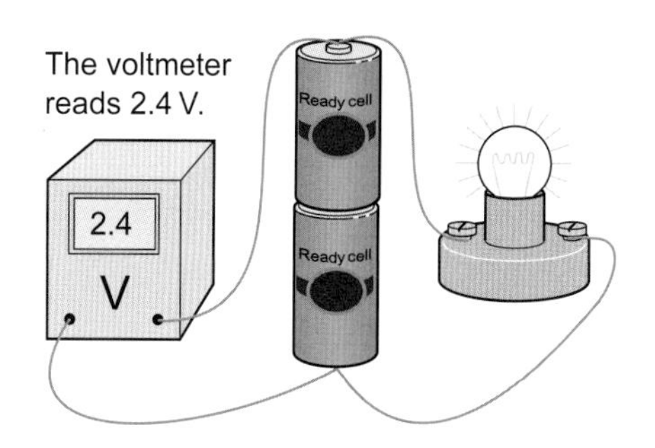

You put a **voltmeter** across a cell or battery to tell you how much energy it can transfer to the current. Cells, batteries and power supplies with a large voltage transfer more energy than those with a smaller voltage.

The voltmeter tells us the number of joules the power supply transfers in a second when the current is 1 A.

In the diagram the cells would transfer 2.4 joules of energy every second if the current to the bulb was 1 A.

Another term for the voltage across something is the **potential difference**.

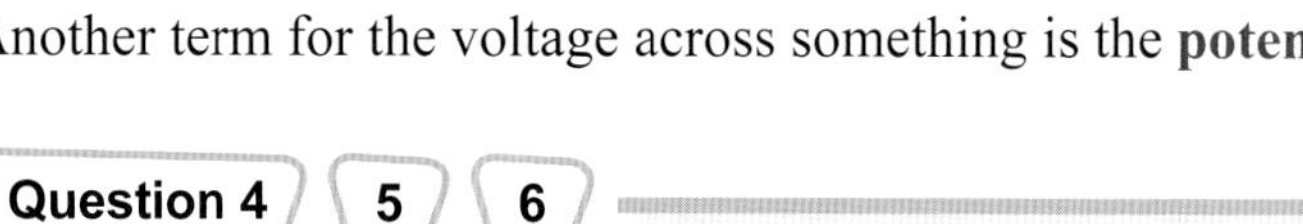

A tour round the circuit

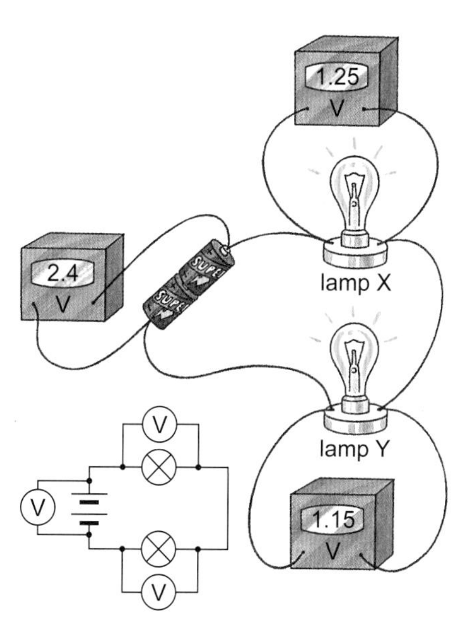

The things in the circuit are called **components**. When the current flows round the circuit, it gains energy from the cell or battery or power supply. This energy is transformed into other types of energy as it goes through the components in the circuit.

In the circuit shown, the components are lamps. The voltage across each bulb shows the change in energy as the current passes through it.

The energy gained by the current in the cells is represented by 2.4 V.
Lamp X transfers energy that gives a drop of voltage of 1.25 V.
Lamp Y has a drop in voltage across it of 1.15 V.

The energy gained by the current in the cells equals the energy transferred out of the circuit in the components.

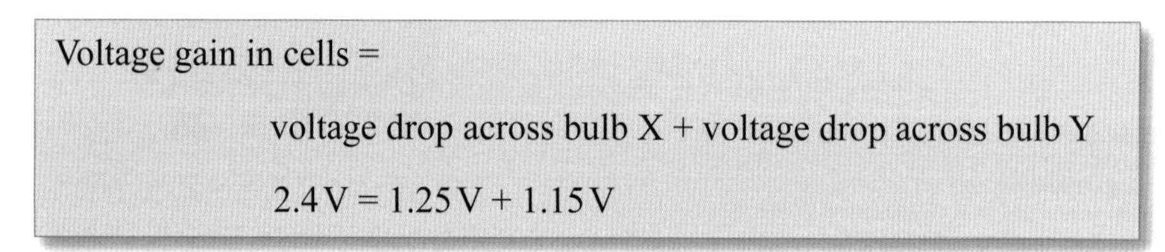

Voltage gain in cells =

voltage drop across bulb X + voltage drop across bulb Y

$2.4\,V = 1.25\,V + 1.15\,V$

High voltages are dangerous

High voltages are dangerous. The cables carrying electricity across the country have to carry a lot of energy, so they have very high voltages. They are put on tall pylons to keep them out of the way of people. Sometimes accidents still happen. The cartoon shows a dangerous situation.

9I.3 Paying for it

You should already know | **Outcomes** | **Keywords**

The way we live today

We use a lot of electricity in the way we live today. The table compares a home in 1909 with one in 2009.

1909	2009
candles	electric light bulbs
open fire	electric cooker
family games and conversation	television
letter	email and telephone

Chemical energy stored in the battery pack changes into electrical energy when the phone is used.

There are also many items that use electricity from batteries or rechargeable cells, such as mobile phones. Our lifestyle today uses a lot of energy in the form of electricity. All this has to be paid for.

Question 1

The power of the mobile phone charger is shown in watts (W) on the rating plate (circled in blue).

How do we know how much energy things use?

Things that transform electricity into other forms of energy are called **appliances**. Different appliances transform different amounts of energy per second.

The energy an appliance transforms per second is called the **power** of the appliance. Power is measured in **watts**, symbol W.

1 W means that something transforms 1 joule of energy every second.

Because some appliances transfer a lot of energy, we also use a bigger unit, the **kilowatt**, symbol kW.

1 kW is 1000 W. 1 kW means that something transfers 1000 joules of energy every second.

A typical jug kettle has a power of 2.2 kW. It transfers 2200 joules per second when it is turned on.

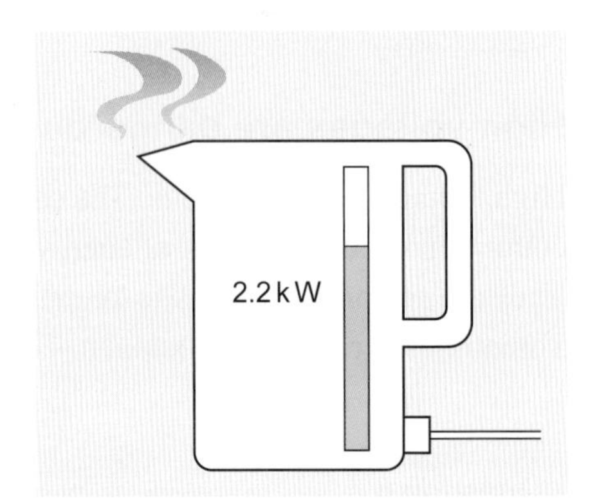

Question 2 **3**

Some things use more energy than others

We use a range of items in the home. Heating appliances usually transfer more energy than appliances that do not heat things. They usually have a higher power rating. The list gives some typical power values.

Appliance	Energy used per second	Power in kW	Power in W
cooker	5000 joules	5 kW	5000 W
TV	500 joules	0.5 kW	500 W
desk lamp	60 joules	0.06 kW	60 W
toaster	1200 joules	1.2 kW	1200 W

Not all the energy transferred by an appliance goes where you want it to go. Some is wasted. How much is wasted depends on the appliance. The diagram shows what happens with a kettle.

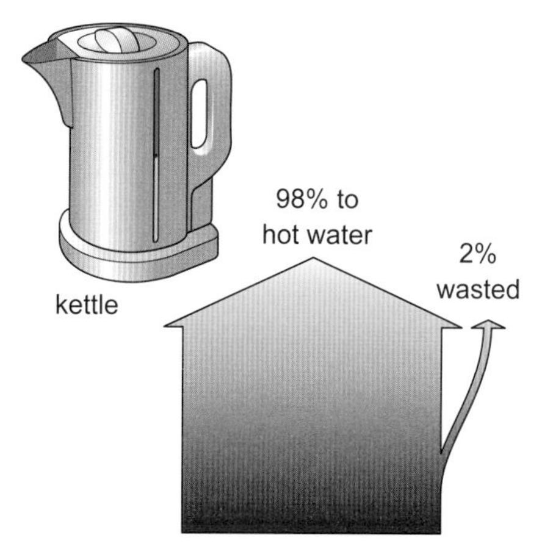

These diagrams show what happens to the energy supplied to the appliance. The arrows show how much energy is turned into useful thermal energy, and how much is wasted.

An old-fashioned electric light bulb has a wire in it that gets hot and wastes a lot of energy in heating up the surroundings.

Energy efficient light bulbs have been developed that do not produce as much heat. They are more efficient because less energy is wasted as thermal energy and so more of the electrical energy is changed into light.

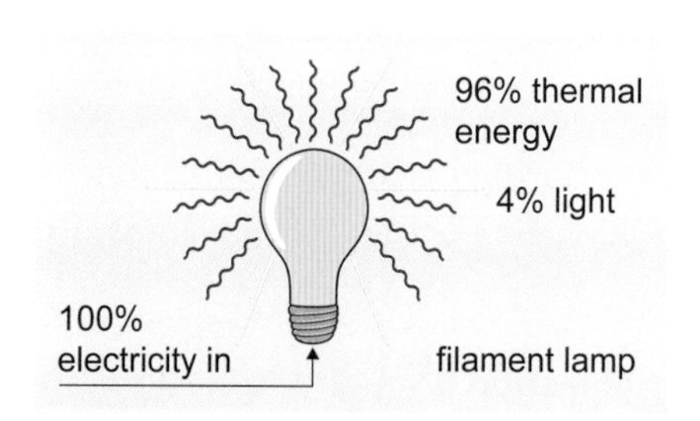

A typical energy saving bulb will use five times less electrical energy to produce the same level of light as an old-fashioned filament bulb.

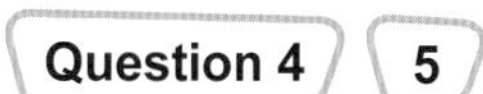

Energy is never destroyed

When energy changes from one form to another, no energy is lost.

You always have the same number of joules at the end as you started with. This idea is called the **conservation of energy**.

The diagram shows what happens to petrol in a car engine when the car is speeding up. About a quarter of the energy (25%) goes into the kinetic (movement) energy of the car. The rest goes into thermal (heat) energy and sound, which are less useful. This type of diagram is called a Sankey diagram.

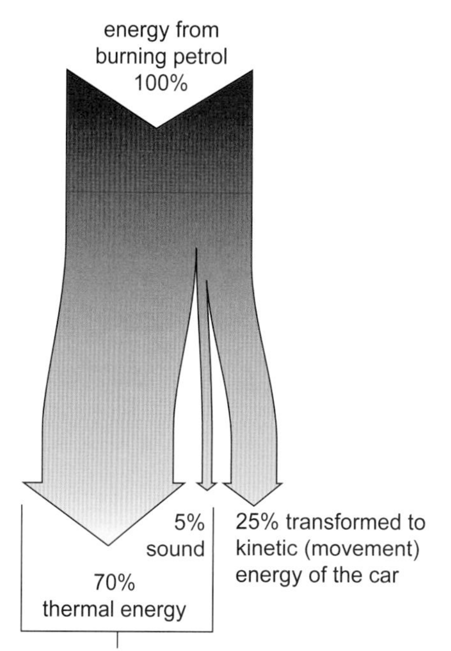

Energy transformations in a car engine.

We say that the energy change is about 25% efficient because 25% of the energy you put in ends up where you need it to be.

You should already know

Outcomes

Keywords

Making electricity on a large scale

Electricity has to be made. One way of doing this is to **generate** large amounts of electricity at a power station and supply it to large areas of the country along a system of wires. To generate electricity we need a source of energy like coal, oil, gas, nuclear fuel, wind or something similar.

Most of our electricity comes from power stations that use **fossil fuels**. This pattern of using fossil fuels is common across the world. Because there is an increase in the use of renewable energy, there are an increasing number of wind turbines supplying energy in Britain. We also have some power stations that use nuclear fuel.

There are two problems with the use of fossil fuels – they are running out and using them is contributing to **global warming**.

Nuclear fuel is an alternative to fossil fuel, but there are dangers from radioactive waste and the radioactivity of the fuel itself.

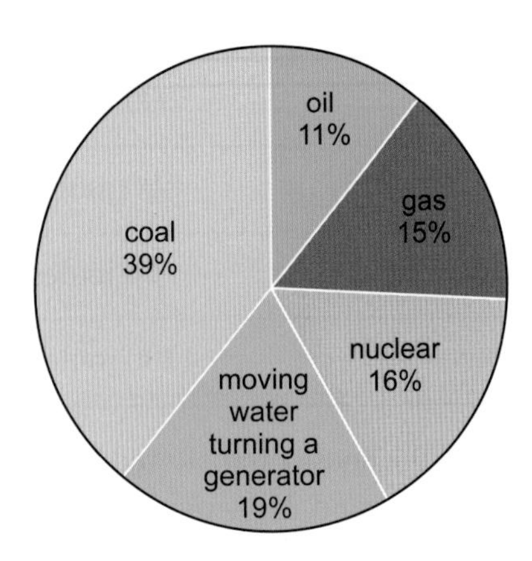

Energy sources used to generate electricity. These figures are for the whole world.

Question 1 2

What happens at the power station?

The fuel is used at a power station to heat water. The hot water produces steam, which turns a **turbine** connected to a **generator**. This produces the electricity when it spins.

coal
oil
gas
fossil fuels
nuclear fuel
(uranium and plutonium)

Fuels we use to generate electricity.

Coal power station or oil power station.

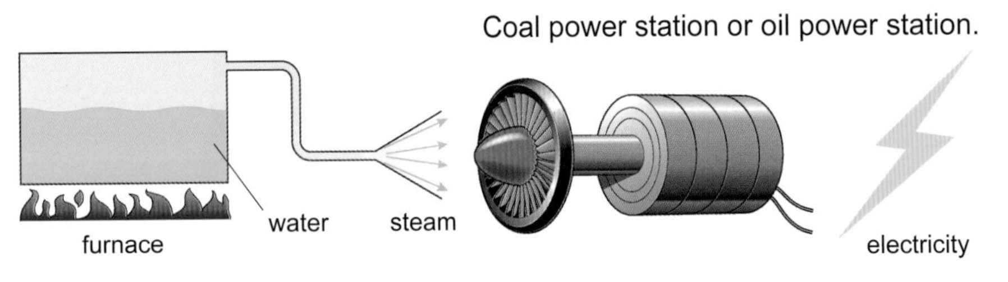

Nuclear power station.

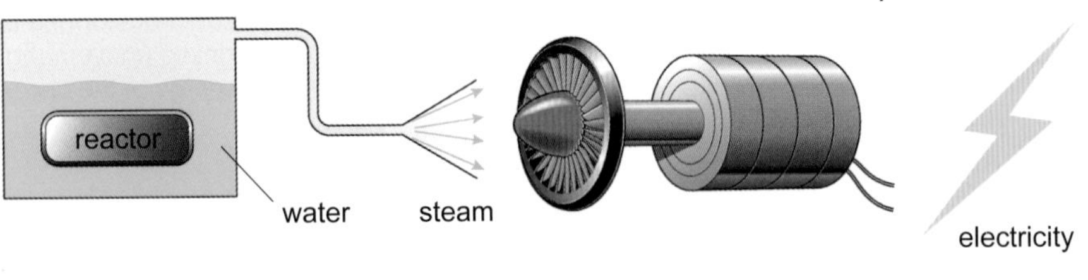

Question 3

What happens in a generator?

Some bicycles have a small version of a power station generator to light their lights. This is also called a **dynamo**. The wheel of the bike turns a coil of wire near a magnet. This generates electricity in the coil, which lights the lights as long as you keep moving.

Some generators have a magnet spinning in a coil of wire. Others use a coil of wire spinning in between magnets. Large generators usually use electromagnets made from large coils of wire with their own power supplies. The basic idea is:

movement of a coil + magnetism = electricity

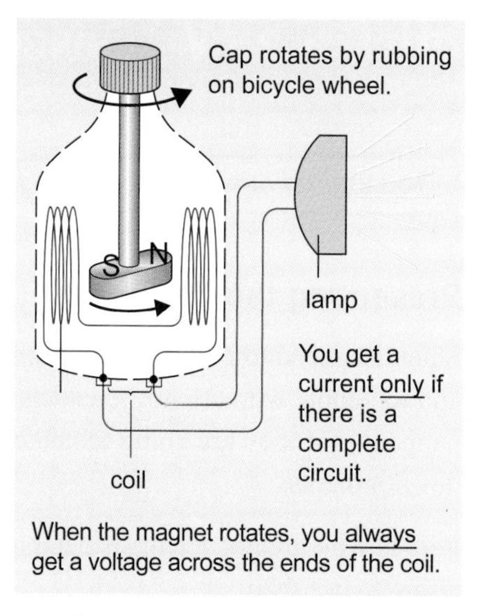

A bicycle generator (sometimes called a dynamo).

Steam is used to make the turbine spin in a power station. The turbine makes the generator rotate to produce the electricity. In wind turbines and in hydroelectric power stations, the turbine is turned directly by moving wind or water. This means the generator is turned without having to heat water. The diagrams show you how.

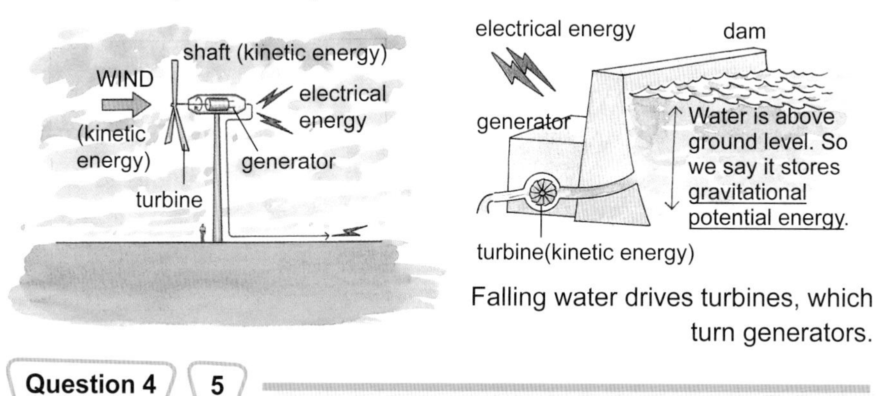

Falling water drives turbines, which turn generators.

Electricity pylons carry very high voltage electricity safely around the country.

Question 4 | 5

How does electricity get to our homes?

Once the electricity has been generated at a power station, it flows into a system of wires called the National Grid.

This is a network of cables on pylons, which carry electricity all around the country. The pylons support the cables at a great height. It is important that low flying aircraft do not go into them.

The wires on the pylons carry electricity at very high voltage. At various places the electricity is carried down from the wires on the pylons to substations. At a substation the voltage of the electricity is reduced to a safer level. It is then passed along wires to homes, shops and offices for use.

Electricity enters your house through a meter that measures how much you use so you can pay for it.

The reading from an electricity meter like this is used to prepare your electricity bill.

Review your work

Summary

9I.HSW Describing energy transfers

You should already know	Outcomes	Keywords

Presenting the information

Gro Harlem Brundtland, former Prime Minister of Norway.

Scientists need to **communicate** things they find out to other scientists and to people who are not scientists. Politicians need to know what is happening if they are going to make decisions on spending and to support developments.

People who are not scientists need to know about developments if they are going to use them.

Gro Brundtland was the Prime Minister of Norway. She was in charge of an international commission in environmental development. It produced an important report about the needs of human beings, the environment and the Earth's resources intended for a wide audience including political leaders.

Sometimes it is easier to communicate ideas in diagrams rather than words. Reports usually use a combination of words and diagrams.

Question 1 

Look at the information in the table and compare it to the chart.

Region	Tonnes of carbon emitted per person per year
Africa	0.6
Asia and Oceania	1.1
Central and South America	1.3
Europe	4.3
Middle East	3
North America	9

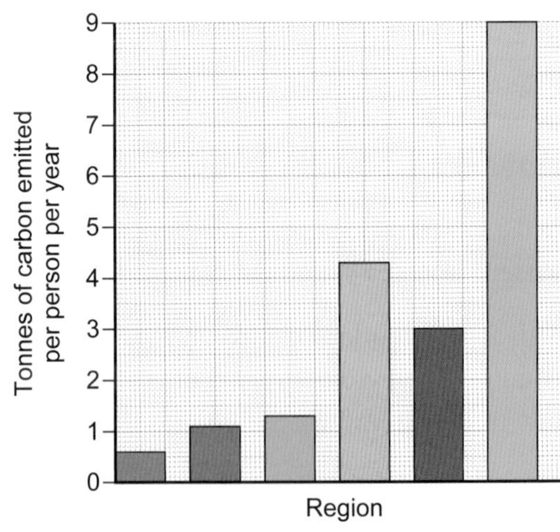

The chart is more **visual** than the table. Some people find visual information easier to understand than text. If you were presenting the information to a group using a computer, using text and visual information would help you to reach more people in your audience.

Other types of chart

The chart on the previous page is called a bar chart. It allows you to make easy comparisons between amounts.

People in North America produce three times as much carbon per person as do people in the Middle East.

Sometimes you want to show information in a different way. The table below compares the different energy sources for electricity. The chart on the right shows how to present it showing how the production of electricity is shared amongst them.

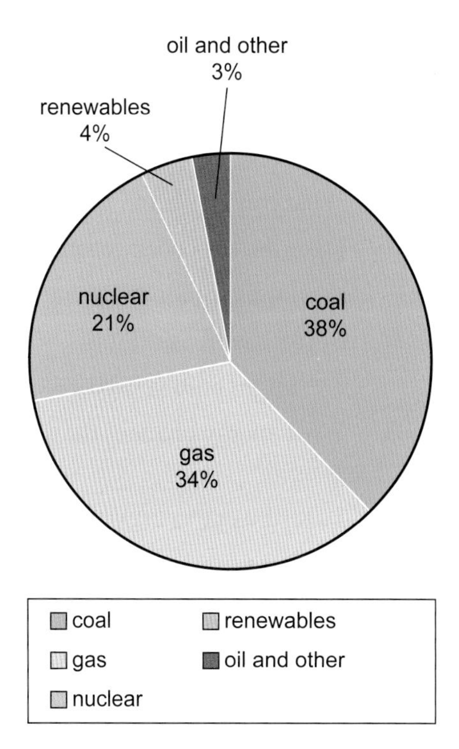

How we generated electricity in the UK in 2005

Energy source	Percentage of electricity produced
coal	38%
gas	34%
nuclear	21%
renewables	4%
oil and other	3%

This type of chart is called a **pie chart**. This is another visual method of presenting data.

Energy transfer diagrams

Energy transfer diagrams are used to show how energy is transferred by an appliance. The width of each arrow is in proportion to the amount of energy that transfers down that route. These diagrams are called **Sankey diagrams**.

The illustration shows two ways of showing the energy transfers in a television set. The data is in this table.

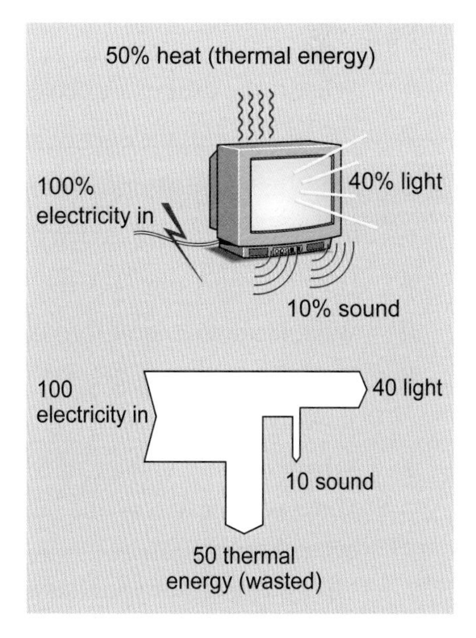

How the television set transfers energy.

Form of energy	Amount transferred by TV set
light	40%
sound	10%
heat	50%

The Sankey diagram makes it easy to see that if you can design the television so it does not produce as much thermal energy, you would need less electricity to run the set.

9I Questions

9I.1

1. What can happen to the energy when an event takes place?
2. Describe the energy changes that take place when:
   - **a** a radio produces a sound;
   - **b** an electric drill turns;
   - **c** a light bulb is used.
3. What energy changes do you think take place when an electric hairdryer is used?
4. Describe one energy change used to heat a room with a solid fuel fire.
5. What type of energy is stored in a fuel?
6. **a** Where did the energy stored in fossil fuels come from originally?
   - **b** What was the name of the process involved in storing it as chemical energy?
7. What does the term 'potential energy' refer to?
8. Give one example of something storing 'gravitational potential energy'.
9. Give two examples of gravitational potential energy transferring to kinetic energy.

9I.2

1. What is the difference between a series circuit and a parallel circuit?
2. Where does all the energy given out by the bulbs come from?
3. **a** What is an electric cell made from?
   - **b** What is the difference between an electrode and an electrolyte?
4. How do you know how much energy a cell can supply to a circuit?
5. What is another term for voltage?
6. How many joules can a 1.5 V cell deliver every second if the current flowing is 1 A?
7. A battery has a voltage of 6 V. It is connected to two bulbs in series. If the voltage across one bulb is 4 V, what is the voltage across the second bulb?
   Give a reason for your answer.
8. Describe two situations where the high voltage of a power line on a pylon can cause an accident.

9I.3

1 Suggest why people living in 2009 use more electricity than people did in 1909.

2 a What is power?

b What are the units used for power?

3 a Explain the difference between watts and kilowatts.

b How many joules does a 3 kW heater transfer to a room every second it is on?
Explain how you worked out your answer.

4 a Which type of appliances usually uses the most energy in the home?

b Give an example that illustrates the difference in the power used by different types of appliance.

5 What is the key feature of an energy saving light bulb that means it is more efficient than an old fashioned filament bulb?

6 What is meant by the conservation of energy?

9I.4

1 What is the difference between the generation of electricity at a power station and the generation of electricity by a wind turbine?

2 Give two problems with using fossil fuels to generate electricity and two problems with the use of nuclear fuel as an alternative.

3 Outline the process used at a power station to generate electricity from fossil fuel.

4 What is the name of the small generator that is sometimes found on a bicycle and how does it work?

5 What is the basic principle used in all electrical generators?

6 How does electricity get from power stations to homes?

9I.HSW

1 Why is it necessary for scientists to be able to communicate their findings?

2 What combination of things do people usually use to present their findings? Suggest a reason for this.

3 Give one advantage of presenting information in a chart rather than a table.

4 a What is the advantage of a pie chart over a table in presenting information about the share of energy produced from different resources?

b Comment on the relative proportions of energy produced from the different resources.

5 Give two different types of chart you might use to present scientific information.

6 a What is a Sankey diagram?

b When might it be useful for presenting information?

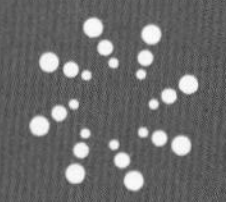

Scientific enquiry

9J.1 What is gravity?

You should already know

Outcomes

Keywords

A force that pulls things together

Two magnets can push apart or pull together. Magnets do not have to be in contact to exert a force on each other. The force of **gravity** is similar. Objects do not have to touch each other to exert a force of gravity. But there are some important difference between the forces of gravity and magnetism.

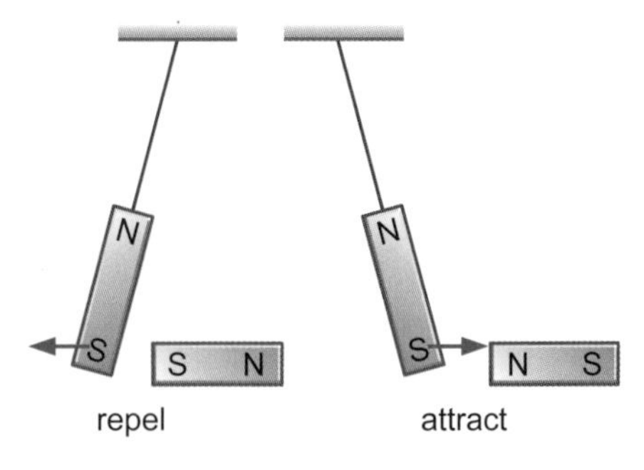

The forces between magnets are quite large.

Gravity is a force that can only pull things together. It never repels. Gravity is a very, very weak force and it acts between all objects. Magnetism only occurs between magnets and a few other substances including iron, nickel and cobalt. Gravity happens because objects have a **mass**. So, everything is affected by the force of gravity. You only notice the force of gravity when one of the objects is very large, like the Earth.

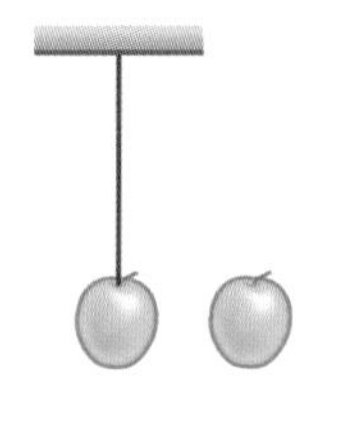

The force of gravity between two small objects is too small to notice.

On the Earth, the force of gravity pulls the object towards the centre of the Earth. It is like having an elastic band stretched between a large, heavy object and a small, light ball. The pull of the band acts on both things, but the effect is to move the small, light ball towards the large, heavy object.

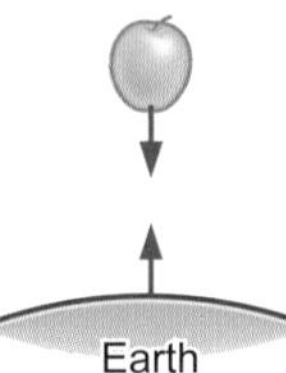

The Earth has a very big mass. So there is a large force of gravity between the Earth and other objects.

Question 1 2

Gravity and weight

The mass of an object tells us how much matter there is in it. An object with a very big mass, like a planet, produces a large force of gravity.

The gravitational pull on a mass by a planet or a moon is often called **weight**. The weight of a small apple on the surface of the Earth is about 1 **newton**. Objects have a smaller weight on the Moon compared to their weight on Earth because the Moon has a smaller mass than the Earth.

The pull of gravity on the Earth is 10 N for every kilogram of mass. This is called the **gravitational field strength**. The gravitational field strength at the Earth's surface is 10 N/kg. The weight of something varies from place to place in the universe because the gravitational field strength varies. The mass does not change.

Mass is how much stuff there is. We measure mass in kilograms (kg).

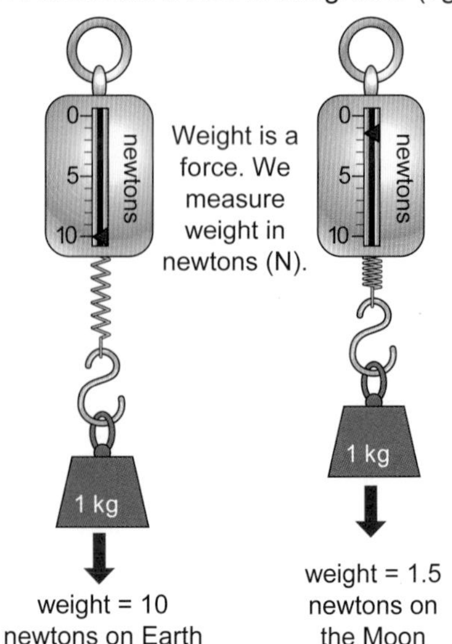

Question 3 4

Gravity and distance

The force of gravity between two things depends on the masses of both objects. It also depends on how far apart the masses are. The force of gravity gets weaker as the objects get further apart.

Isaac Newton used this idea in 1665 to explain how the Solar System worked and why tides happen on the Earth. His ideas are still used today.

Gravity gets weaker as you move away from something across space, but it never disappears altogether.

The force of gravity from the Moon felt on the Earth's surface is weak because the Moon is 250 000 miles away from the Earth. It is still enough to affect the water on the surface of the Earth. The position of the Moon causes the water on the surface of the Earth to bulge. As the Earth spins, the bulge occurs at different places and we notice it as high and low tide.

The force of gravity from the Sun is only about half as big as the Moon's at the Earth's surface. This is because the Sun is a lot further away than the Moon. Nevertheless, the force of gravity from the Sun adds to the force from the Moon. Twice every month, the Sun and the Moon are in line. When this happens, the gravitational forces act together and we get a higher than normal tide, called a spring tide.

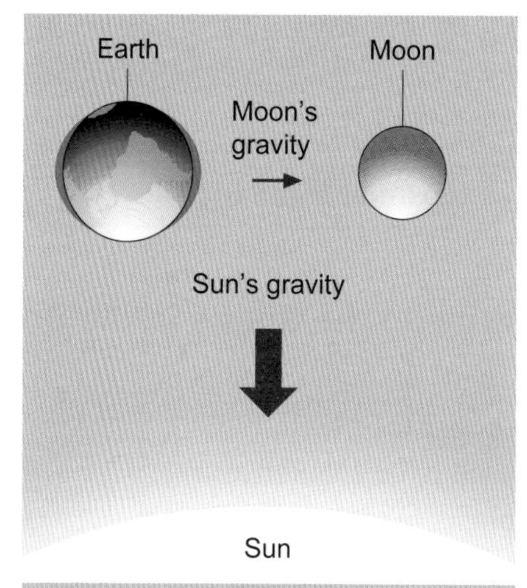

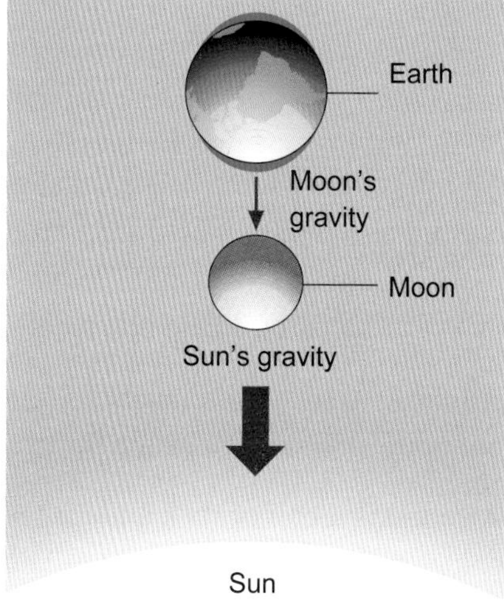

When the Sun and Moon are in line the tides are higher than usual.

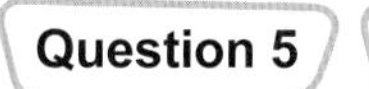

Gravity and space travel

Rockets taking off from Earth can burn 13 tonnes of fuel in a second. This creates the force needed to move away from the Earth's gravity. As the rocket moves further from the surface of the Earth, the gravity that the rocket feels gets less, and so less force is needed to keep it moving away from the surface of the Earth.

Eventually, a rocket travelling between the Earth and the Moon will reach a point where the gravitational pull from the Moon is stronger than the pull from the Earth. At this point, it will start to fall towards the Moon rather than have to pull away from the Earth.

When the space module leaves the Moon to return to Earth, a lot less fuel is needed because the Moon's gravity is a lot smaller than the Earth's. So the force needed to pull away from the Moon's gravity will not be as large as that needed to pull away from the Earth. The point where the Earth's gravity starts to pull the space module towards it is also closer to the Moon than the Earth, which again means that less fuel is needed for the return journey.

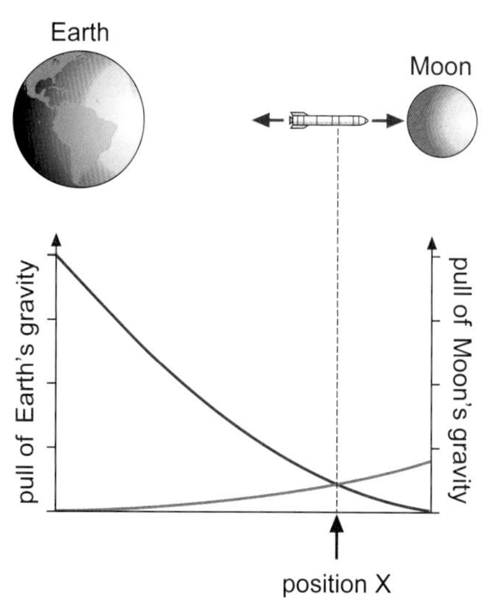

Position X shows the point where the Earth's gravity balances the Moon's gravity.

Check your progress

9J.2 The Solar System

You should already know | Outcomes | Keywords

Why do the things in the Solar System go round the Sun?

There are eight **planets** in the Solar System. Pluto used to be counted as a planet but it is now thought to be more like an **asteroid**.

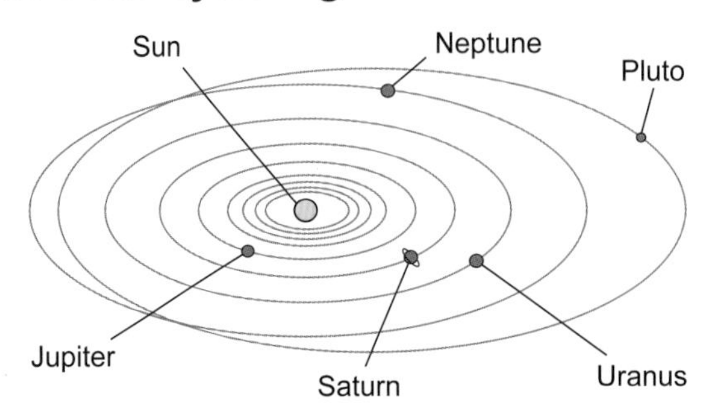

Comet West.

An asteroid is a lump of rock that orbits the Sun but is too small to be classed as a planet. There are a large number of asteroids in a belt orbiting the Sun between Mars and Jupiter. Some people think they are the remains of a planet that broke up into small pieces a long time ago.

All the asteroids and planets go round the Sun in oval orbits. The orbits are not quite circular. The name for their shape is an **ellipse**.

A **comet** is a ball of rock and ice that also orbits the Sun. The orbit of a comet is a very extreme ellipse. The Sun is close to one end of the orbit which means that a comet only passes near the Sun once every hundred years or so and spends most of its time a long way from the Sun. When we make observations from the Earth without a telescope, we only see the same comet every hundred years or so.

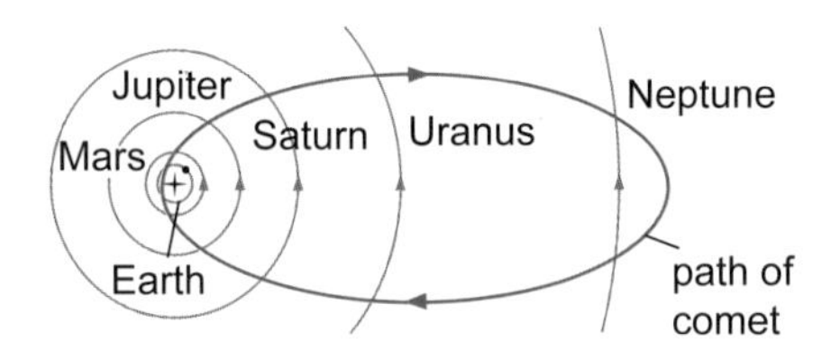

Comets have very large, very elliptical orbits. They can take tens, or even thousands, of years to make one orbit. During the most distant parts of their orbits, comets move more slowly.

All the planets, asteroids and comets go round the Sun because of the pull of gravity between them and the Sun.

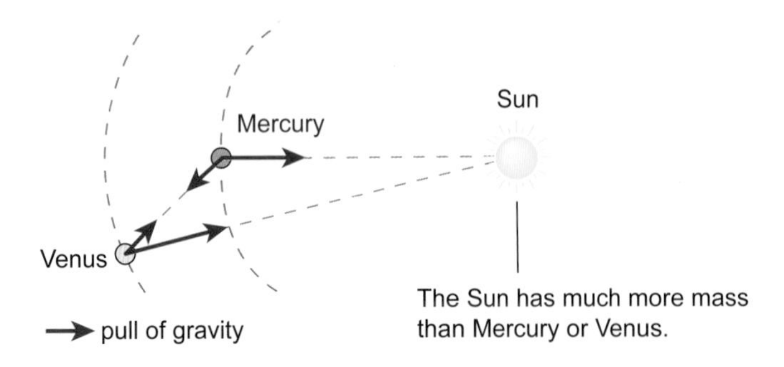

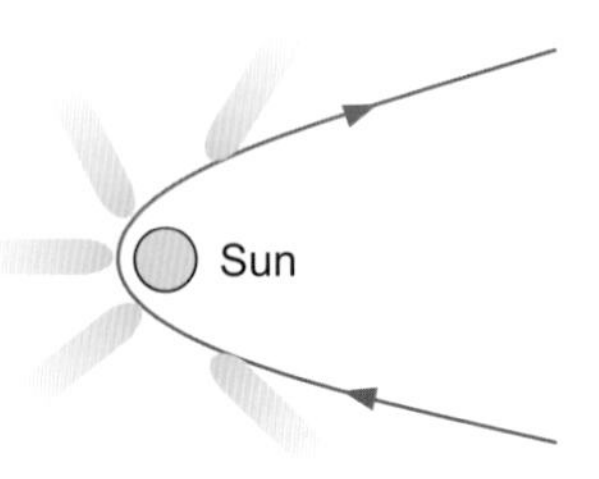

You can see comets when they are closest to the Sun. Energy from the Sun melts the ice, and the 'solar wind' makes the vapour into a tail.

The Sun is so massive compared to everything else that there is an enormous pull between it and every other object in the Solar System.

The Sun is 330 000 times heavier than the Earth!

Question 1 | 2 | 3

Why don't the planets fall into the Sun?

The Earth has a speed through space along the path of its orbit of about 30 kilometres per second. The gravitational pull of the Sun is constantly pulling it inwards. If the Earth moved more slowly it would be pulled into the Sun because of the Sun's gravitational pull.

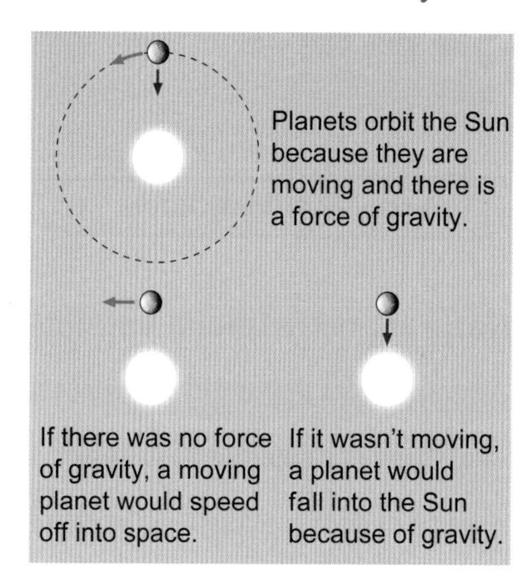

Imagine swinging an object round on a piece of string.

You are like the Sun. The piece of string is pulling on the object like the pull of gravity from the Sun pulls on the Earth.

Because of the pull of the string, the object does not go in a straight line – it is pulled into a circular path. The pull of gravity stops the Earth moving in a straight line but the speed of the Earth is just right to keep it moving in an orbit.

For the Earth, the force of gravity pulls it into an elliptical path. You could model this with an object if the string you used was elastic.

Question 4

Why doesn't the Sun go round the planets?

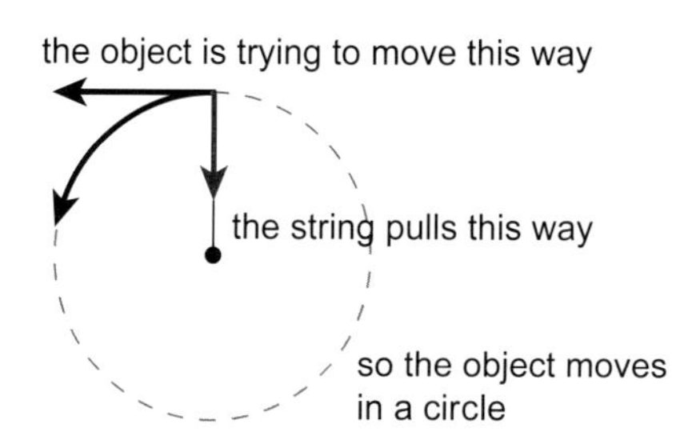

The person swinging the object round on a string is pulling on the string with the same size of force as the object exerts in pulling away. They each feel a pull from the string.

The object goes round the person rather than the person going round the object because the person is a lot heavier.

The same idea happens with the Sun and a planet like the Earth. Because the Sun is so massive compared to the planets, asteroids and comets, the planets, asteroids and comets go round the Sun and not the other way round.

Question 5

You can't always believe what you see!

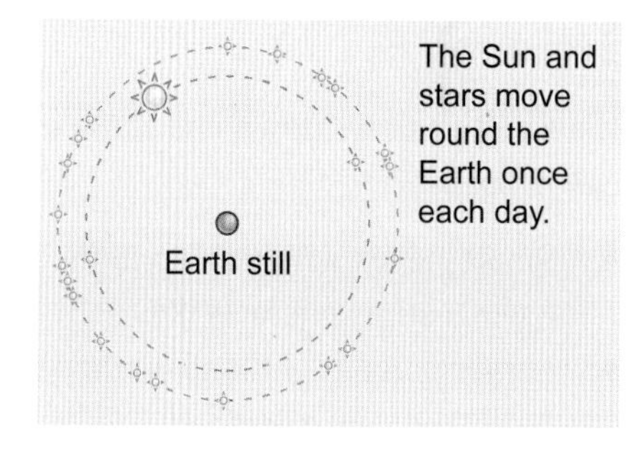

What people used to think happens.

If you look at the sky over the course of a day, it looks like the Sun is moving around the Earth. The explanation is that the Earth spins on its axis once every 24 hours. People thought this was a crazy explanation at first. They thought the spin would make people fall off the Earth.

Newton explained why we do not fall off a spinning Earth. The objects on the Earth are moving with the Earth at the same speed as it is spinning. This is why we don't fall off. It is like pouring a drink in a train or on an aircraft. If you are going at a steady speed, the drink does not fly backwards because it is moving at the same speed as the train or aircraft.

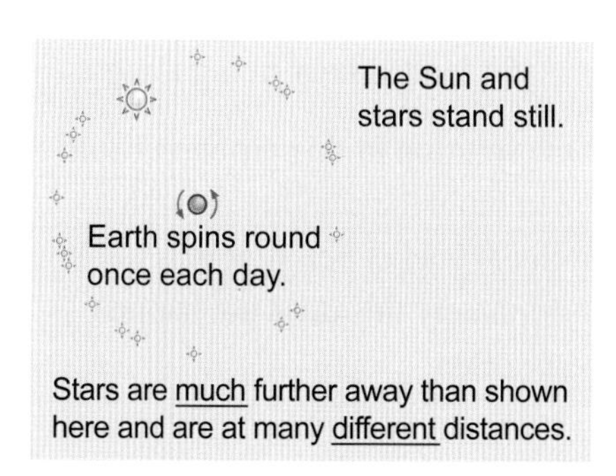

What we now think happens.

Question 6

9J.3 Satellites

You should already know

Outcomes

Keywords

Natural and artificial satellites

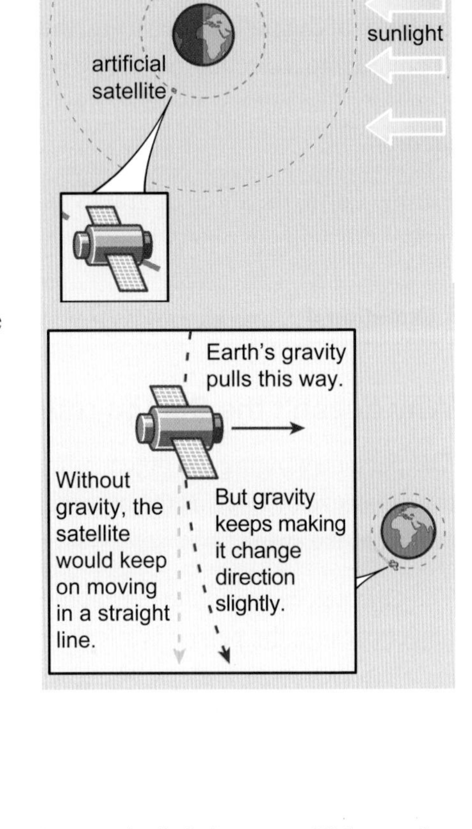

A small object that orbits a planet is an example of a **satellite**.

The Moon is a satellite of the Earth. It is called a **natural satellite**.

A lot of the other planets in the Solar System have natural satellites or moons. Mars has two moons, which we call Phobos and Deimos.

Jupiter has at least 16 natural satellites or moons. There are four large ones, which Galileo observed going round Jupiter with one of the first telescopes in 1610.

Humans put artificial satellites into orbit around the Earth. Satellites orbit the Earth in different ways depending upon their purpose. The Moon is large enough to see with the naked eye. Artificial satellites can be seen with a telescope. We see satellites because sunlight reflects off them.

All satellites are kept in orbit by the pull of gravity.

The satellite has to be moving along its orbit at just the right speed so that it stays on its orbit. If it speeds up, it will move away from the Earth. If it slows down, the pull of gravity will cause it to spiral into the Earth.

Satellites orbit above the atmosphere so drag from the air does not slow them down.

If a satellite slows down and starts to spiral into the Earth, several things can happen.

- Small pieces will burn up in the Earth's atmosphere as they fall.
- Larger parts might land on the Earth and cause damage.
- A satellite sometimes contains material which would be dangerous if it landed near living things.

One solution to this problem could be to blow the satellite up with a military rocket before it gets too low. Fortunately this does not happen very often.

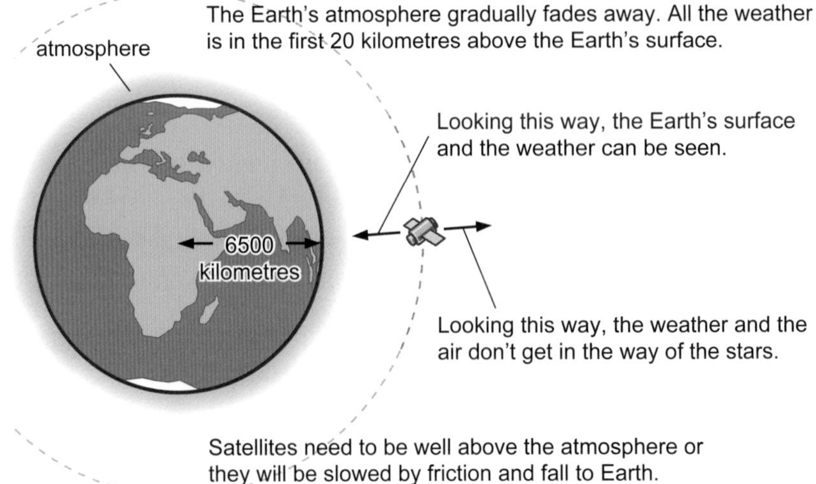

Question 1  3

Satellites for observation

Astronomers have difficulty getting a good view of the stars from the Earth for several reasons. The main ones are:

- Clouds get in the way.
- Light pollution from cities reduces the contrast.
- Air pollution from industry reduces the quality of view.

Telescopes are put on the top of remote mountains to avoid most of these problems.

You can put a telescope on a satellite to get very clear pictures. One of the most famous telescopes on a satellite orbiting the Earth is called Hubble. You can find amazing images of space from Hubble on the internet.

Astronomers have problems trying to see the stars.

Question 4

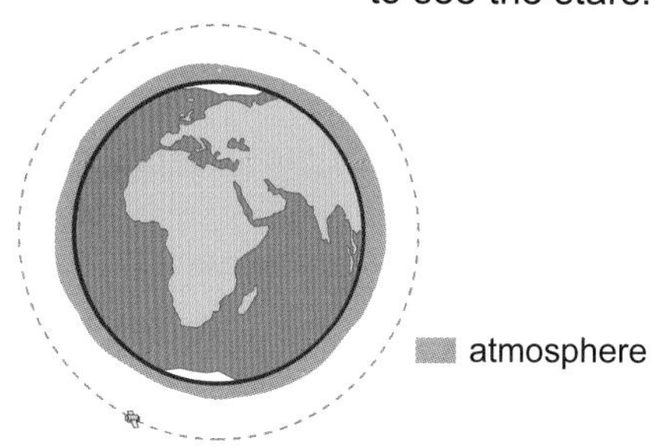

The best place to put a telescope is on a satellite.

Looking down

As well as looking into space, satellites can be used for telescopes that observe the Earth's surface. These have many uses.

Many observation satellites travel fast and quite close to the Earth. They complete several orbits every day, taking photographs of the Earth's surface as the Earth spins under them.

Photographs of the Earth's surface are useful to monitor the weather. They are also used to gather military intelligence.

You can see a satellite photograph of the place where you live by typing your postcode into a map site on the internet and selecting the satellite option.

Satellite image of weather.

Satellite image of fields showing what crops are growing.

Question 5

What is a geostationary orbit?

Some satellites take 24 hours for one complete orbit. They travel in the same direction as the Earth spins. They always appear at the same point in the sky. We call this a **geostationary orbit**. We use this type of satellite for TV broadcasts.

The diagram shows how a signal from a broadcasting station reaches a house with a satellite dish. A geostationary satellite is always above the same part of the Earth's surface so the dish on the side of the house stays fixed, pointing at the same spot in the sky. By using the satellite, it is also easy to send signals around the curve of the Earth's surface.

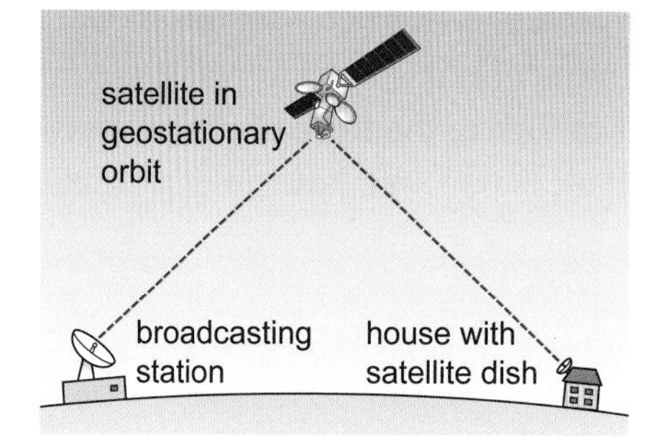

Question 6

Review your work

Summary ➡

9J.HSW Models and predictions

You should already know | **Outcomes** | **Keywords**

Using a model

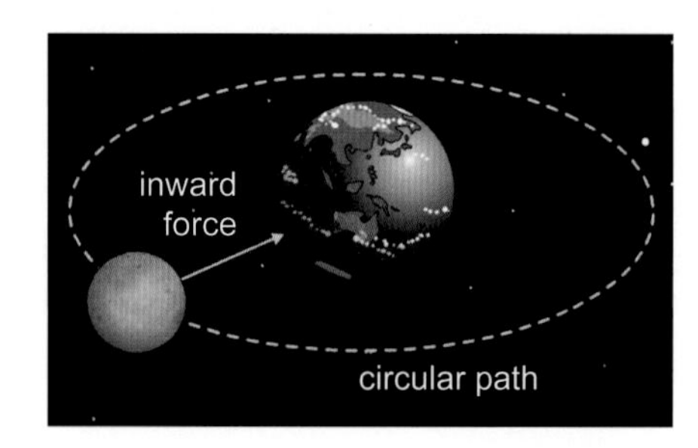

Scientists use models to help them understand events. A **model** is a simple version of the event the scientist is trying to understand. The model will have some things in common with the event you are studying. It will also have some differences.

The Moon goes round the Earth in a circular orbit. It is pulled by the force of gravity. A rubber bung on a string is a model of this.

- The tension of the string on the bung is like the pull of gravity on the Moon.
- The bung is like the Moon.
- The person swinging the bung round is like the Earth.

This model helps answer the question, 'What would happen if the force of gravity between the Earth and the Moon was suddenly switched off?'

This would be like the string breaking.

The diagram shows what happens to the bung if the string breaks.

The bung flies off on a straight line.

If gravity was switched off, the Moon would travel off into space along a straight line.

The problem with any model is that there are differences between the model and the real world. The force on the bung is provided by a string, which is in contact with the bung. The force of gravity on the Moon does not need anything in contact between the Earth and the Moon. Gravity is a force that acts across empty space.

Another difference with this model is that the size of the tension in the string depends upon how fast the person swings their end of the string. The size of the force of gravity between the Earth and Moon depends on their masses and how far apart they are.

← = force (tension in string)

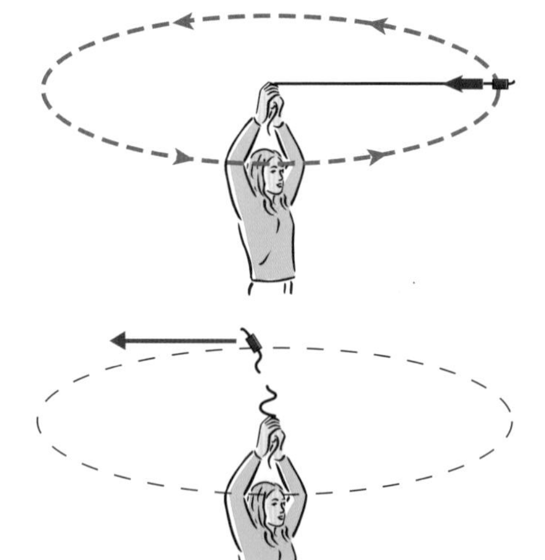

If the string breaks, the bung continues to move as shown. We say that it flies off at a tangent (to the circle).

Question 1  2 3

Making predictions from other data

Sometimes in science you use data to make a **prediction** about something you do not know.

The table shows the temperature of different planets and their distances from the Sun.

Planet	Distance from Sun (million km)	Surface temperature (°C)
Mercury	60	480
Venus	110	350
Earth	150	20
Mars	230	–23
Jupiter	780	–150
Saturn	1430	–180
Uranus	2870	–
Neptune	4500	–220
Pluto	5900	–230

The surface temperature gets lower as you get further away from the Sun. If you plot a graph of the data you can make an estimate of the temperature on Uranus by drawing a smooth curve between the points.

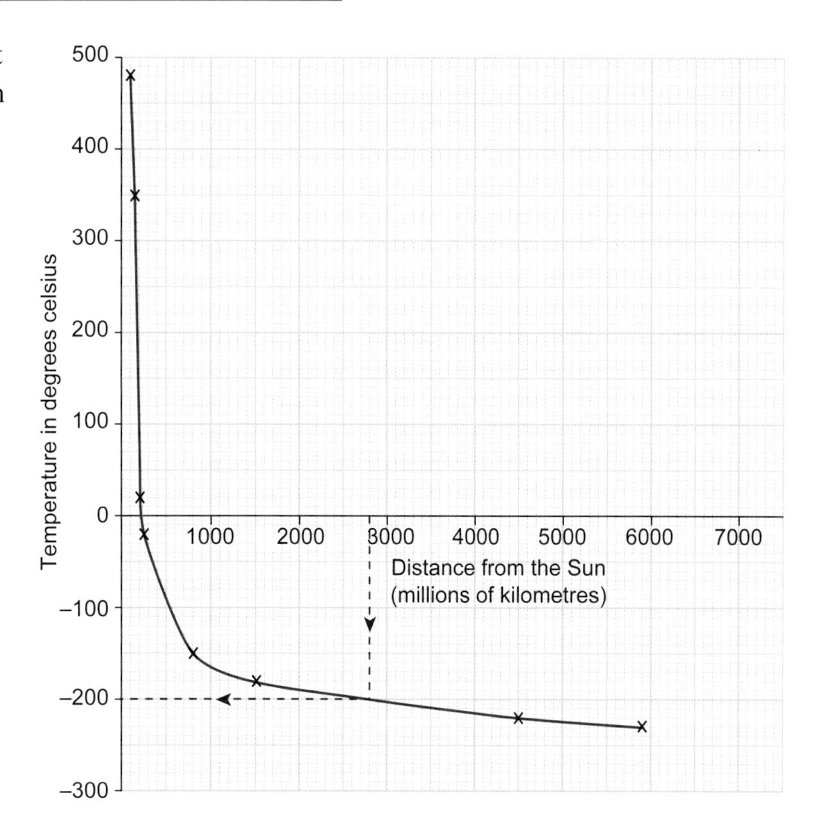

If you draw a line on the graph at the distance for Uranus, you can use it to predict the temperature on Uranus as –198 °C.

Question 4  5  6

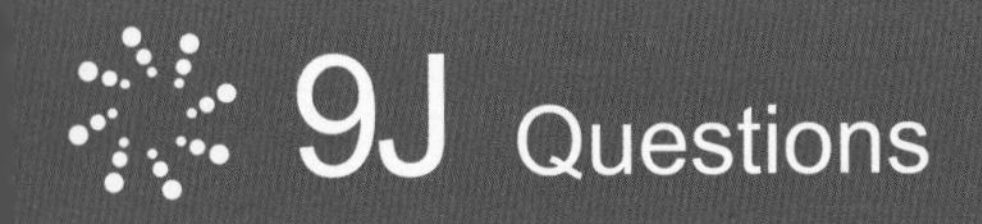

9J Questions

9J.1

1 Describe the similarities and differences between the forces of magnetism and gravity.

2 a What property of objects causes the force of gravity?

b Why do all objects feel the force of gravity?

3 What is the name for the pull of gravity on an object at the Earth's surface and why is it larger than it is at the Moon's surface?

4 a What is the gravitational field strength at the Earth's surface?

b What does it tell you about the weight of a 3 kg mass?

5 What are the two things that affect the size of the force of gravity between two objects?

6 Describe one effect of the Moon's gravity on the Earth.

7 Explain why a spacecraft uses a lot less fuel returning to Earth from a trip to the Moon than it did to get there.

9J.2

1 What are the differences between an asteroid, a planet and a comet?

2 What is the name for the shape of the path taken round the Sun by planets, asteroids and comets?

3 How does the mass of the Sun compare to the mass of the Earth?

4 How does the force of gravity between the Sun and the Earth make the Earth move in an orbit?

5 Why doesn't the Sun orbit the Earth?

6 Why don't we fall off the Earth if it is spinning as well as moving through space at 30 km/s?

9J.3

1 a What is the difference between an artificial satellite and a natural satellite?

b Give one example of a natural satellite.

2 Why can satellites be seen through a telescope?

3 a Why must satellites orbit above the atmosphere?

b What would be the effect of putting a satellite into orbit in the atmosphere?

4 a What are the advantages of mounting a telescope like Hubble on a satellite rather than on the surface of the Earth?

b Suggest three disadvantages to mounting a telescope on a satellite.

5 Give two uses of satellite observations of the Earth's surface.

6 a What is a geostationary orbit?

b Why is it useful for communications systems to use a geostationary satellite?

9J.HSW

1 What is meant by the term 'model' in science?

2 Describe a model we could use for the Moon orbiting the Earth.

3 Describe the important differences between the model of the Moon orbiting the Earth and the actual situation it is being used to describe.

4 a What happens to the surface temperatures of the planets as you get further away from the Sun?

b Suggest a reason for this.

5 Describe how the graph is used to estimate the temperature on Uranus.

6 a Use the graph to predict the distance from the Sun of a planet where the temperature on the surface is –100 °C.

b If this planet existed, where would it fit in the order of existing planets? Give a reason for your answer.

9K.1 Describing how fast something moves

You should already know | Outcomes | Keywords

All about speed

The car and the lorry shown in the diagram go past the lamp-post at the same time.

The car and the lorry pass the lamp-post at the same time.

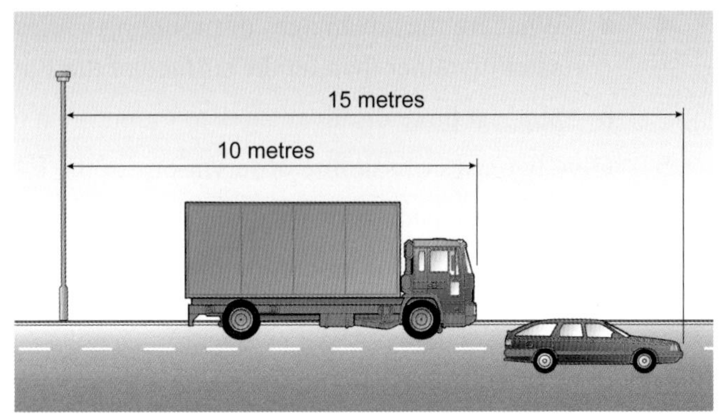

One second later.

One second later the car has travelled 15 metres.
The lorry has travelled 10 metres in the same time.

We say that the **speed** of the car is 15 metres per second.

The speed of the lorry is 10 metres per second.

The car has a higher speed. It goes further in the same time.

The speed describes how far something goes in a certain time. The table shows you some different **units** of speed.

Speed unit	Name for the unit	What the unit means
m/s	metres per second	how many metres the object travels in a second
km/h	kilometres per hour	how many kilometres you travel in one hour
m.p.h.	miles per hour	how many miles you travel in one hour

A speed of 15 metres per second is written as 15 m/s.

Question 1 | 2

Using the different units

In science, the unit of speed is usually metres per second. This makes it easier to make a fair comparison between different speeds. Different units for speed like miles per hour, kilometres per hour are used in the everyday world.

Speed = 30 m/s.

Speed = 20 m/s.

Road signs in Britain display a number but do not show the unit. The unit for British road signs is miles per hour.

Similar signs in France show speed limits in kilometres per hour.

The table gives some examples of the same speed in three different units.

Speed in m/s	Speed in km/h	Speed in m.p.h.
13.3	48	30
17.8	64	40
31.1	112	70

Question 3

Measuring the speed

Cars have a speedometer on the dashboard so the driver can see what speed the car is moving at.

Speed cameras and traffic police measure the speed of a moving car using a speed gun. The gun sends out a burst of microwaves. Some of these reflect off the moving car back to the gun. The reflected waves are affected by the movement of the car, and a computer in the gun can calculate the car's speed.

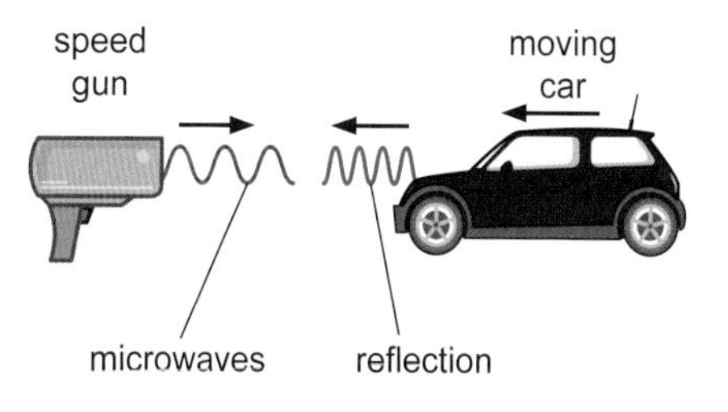

A speed gun detects speed with reflected microwaves.

People who study the weather use an instrument called an **anemometer** to measure how fast the wind is blowing.

You sometimes see an anemometer fixed on a bridge to measure when the wind is too strong for high-sided vehicles.

Measuring a high wind speed with an anemometer.

Question 4 | 5

9K.2 Working out the speed

You should already know | Outcomes | Keywords

Timing things

To compare the speeds of runners in a race, you only need to measure the time they take. This is because everyone in the race runs the same distance.

The timing method shown in the picture shows what happened about 50 years ago. The timekeeper started a watch when he saw the smoke from the starting pistol. The runners set off when they heard the sound of the starting pistol. The timekeeper stopped the watch when the first runner went past. You needed separate timekeepers for second and third places. One problem with this was the error caused by the reaction times of the people acting as timekeepers.

This method is not very accurate. Nowadays the timer is started electronically by a link to the starting gun. There is also a camera system at the finish to register all the athletes' times even when they are very close together. This is sometimes called a 'photo finish'.

Question 1 | 2

Calculating the speed

To work out the speed of something you need to know:

- how far it travels (the distance);
- the time it takes to travel the distance.

You then work out the speed by using this **formula**.

speed = distance travelled ÷ time taken

The speed of the swift in the diagram is worked out like this.

speed = distance travelled ÷ time taken

speed = 240 m ÷ 6 s

The speed of the swift is 40 m/s.

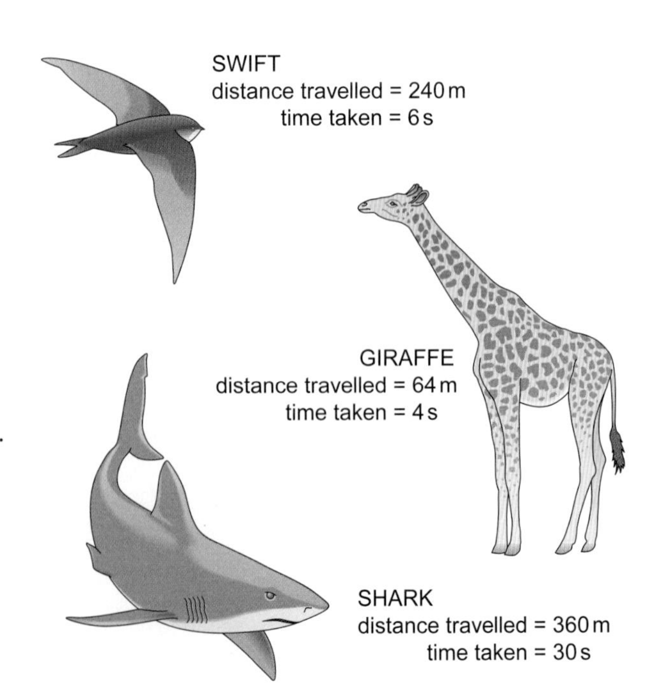

Question 3 | 4 

Other versions of the speed formula

Sometimes you know the speed and the time and you want to work out the distance travelled.

A walker might travel at 6 kilometres per hour. If she walks for one and a half hours, we find the distance she walks with this formula.

distance = speed × time
distance = 6 km/h × 1.5 h
distance = 9 km

Sometimes you know the speed and the distance and want to work out the time. The train from London to Paris goes at about 160 km/h. If it travelled at that speed all the time you could find how long it takes to travel 240 km like this.

time = distance ÷ speed
time = 240 km ÷ 160 km/h
time = 1.5 hours

Eurostar travelling from London to Paris.

Question 5 **6**

Average speed

Things do not usually move at the same speed all the time.

The walker will stop from time to time. She will walk a bit more slowly uphill and go a bit faster downhill. When we say she has an **average speed** of 6 km/h, we mean that in one hour she travels 6 km. This does not mean she walks at the same speed all the time. Sometimes she goes faster, and sometimes she goes more slowly.

The speed of a car will vary depending on the road conditions and traffic. You calculate the average speed for a car journey by dividing the total distance travelled by the total time taken.

The formula could be written like this.

average speed = total distance travelled ÷ total time taken

The clock and milometer from a car shown in the diagram give a total distance travelled of 120 miles and a total time of 3 hours. So the average speed is calculated like this.

average speed = total distance travelled ÷ total time taken
average speed = 120 miles ÷ 3 hours
average speed = 40 miles per hour

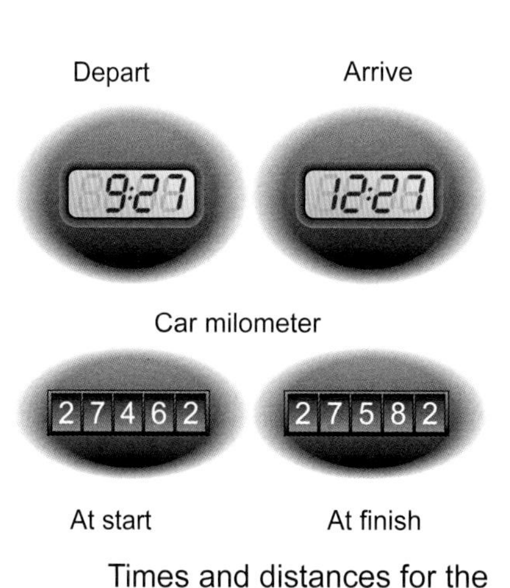

Times and distances for the car journey.

Question 7

9K.3 Force and speed

You should already know | Outcomes | Keywords

Avoiding friction

Friction is a force that makes it harder for things to slide past each other. Friction is a force that slows moving objects down. If you want to study how something moves under the influence of a force, it is a good idea to reduce the friction as much as possible.

To make the force of friction very small we can use a special piece of equipment called an air track. It works in a similar way to a hovercraft. It produces a cushion of air that the moving glider can float on.

When the glider moves, the only friction force is the drag from the air, which is so small it can be ignored.

An air track is a tube with a triangular cross-section.

Air is pumped into the tube and emerges through tiny holes in the side. A glider on the tube floats on the air that comes out of the holes. You can do simple experiments to investigate force and motion using the glider and not worry about friction because it is so low.

A skater gliding across very smooth ice is about as close as you get in the everyday world to movement without friction.

Question 1

Making time measurements

One way to time the motion of the glider is to use a beam of light passing between two tubes, called a **light gate**. If something blocks the light in the gate, a computer can be used to sense this.

When the glider moves through a light gate, the computer times how long the light is blocked for. The computer calculates the speed of the glider from this measurement.

The diagram shows an experiment where the speed of the glider is measured in two different places along the air track.

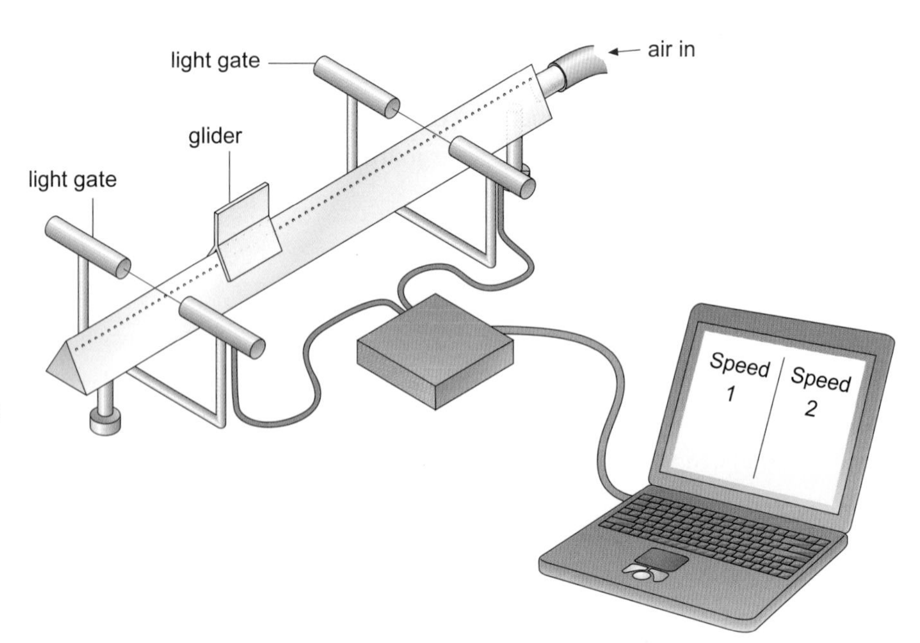

An air track.

Question 2

Applying a force

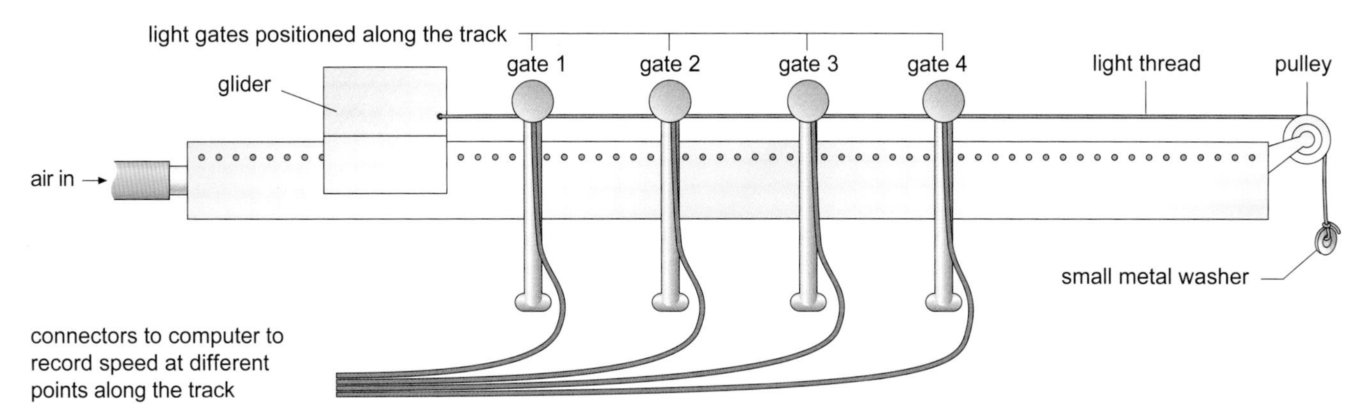

There are three forces acting on the glider in the diagram above:

- the pull of the thread;
- the weight of the glider;
- the push of the air from the air track on the glider.

The weight of the glider is balanced by the push of the air. The pull of the thread is not balanced by anything. We call it an **unbalanced force**.

The results show that the glider is speeding up.

Gate	Speed
1	17 cm/s
2	22 cm/s
3	26 cm/s
4	30 cm/s

Speeding up is a type of **acceleration**. An unbalanced force causes an object that is free to move to accelerate.

Question 3 **4**

What else can an unbalanced force do?

An unbalanced force can change the **direction** something is moving in.

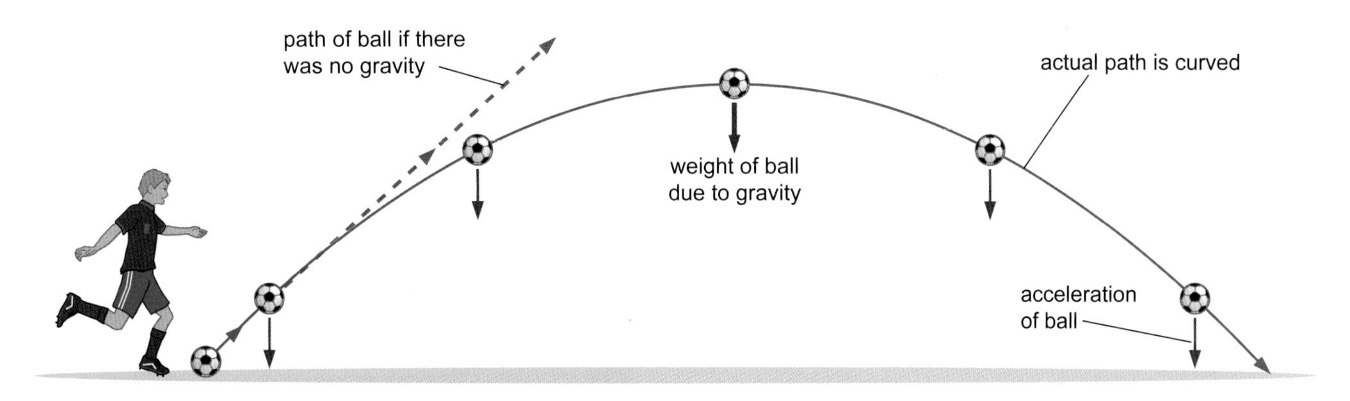

The force of gravity changes the direction of the motion.

Changing direction is another type of acceleration. When we say a force causes acceleration, we mean it can cause something to change its speed, change its direction or do both.

Question 5

9K.4 Slowing things down

You should already know | Outcomes | Keywords

Slowing things down

A force causes acceleration. This means it can change the speed of something or the direction it is moving in or both.

The diagram shows a model car rolling down a slope.

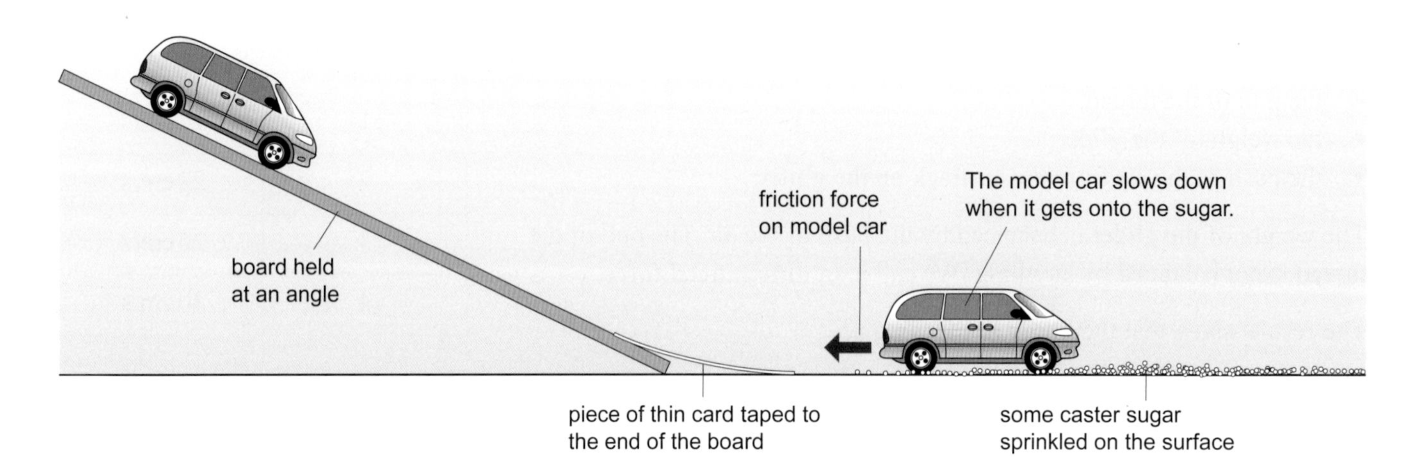

When the car is on the slope, the force of gravity causes the car to speed up as it rolls down the slope.

When the car reaches the thin card, the force of the card on the wheels of the car causes it to change its direction from going down at an angle to travelling along the flat.

When the car reaches the sugar, the friction force from the sugar causes the car to slow down. This change in speed is another type of acceleration.

We use **friction** to slow things down as a safety feature in many situations. You sometimes see gravel and sand tracks at the side of a road at the foot of a steep hill. These are called escape lanes and vehicles going too fast can turn onto them to stop. The gravel and sand will slow down a car that has speeded up too much coming down the hill like the sugar slows down the model car.

A track of gravel is used at the foot of steep hills to stop cars and lorries that are out of control.

Question 1 | 2

Keeping moving

The driving force from the wheels speeds a car up.

As the car speeds up, friction from the air increases until the total friction force balances the driving force from the wheels. The friction force from the air is also called 'drag' or 'air resistance'. The diagrams show three different situations for the forces acting on a car.

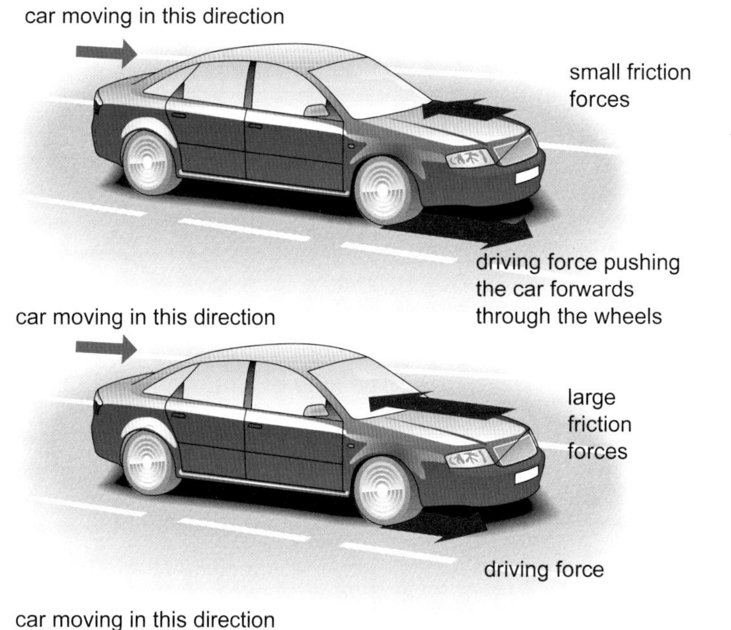

When the driving force is bigger than the friction force, the car speeds up.

When the driving force is smaller than the friction force, the car slows down.

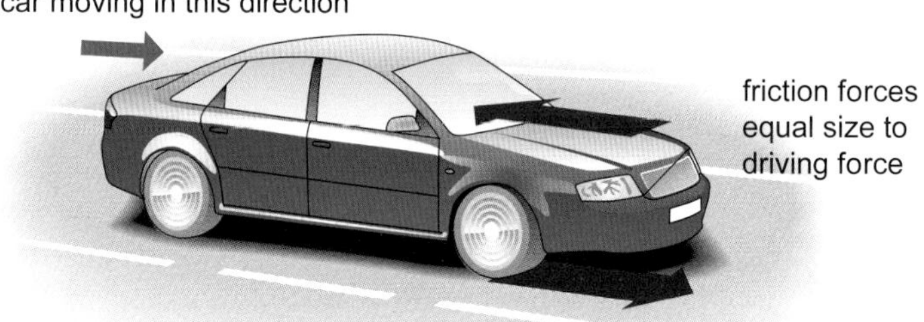

When the driving force equals the friction force, the car goes at a constant speed.

The cyclist and canoeist are both slowing down. In both cases the driving force is zero so the friction force is unbalanced. The fish is moving at a constant speed because the driving force from its tail is equal and opposite to the friction force from the water.

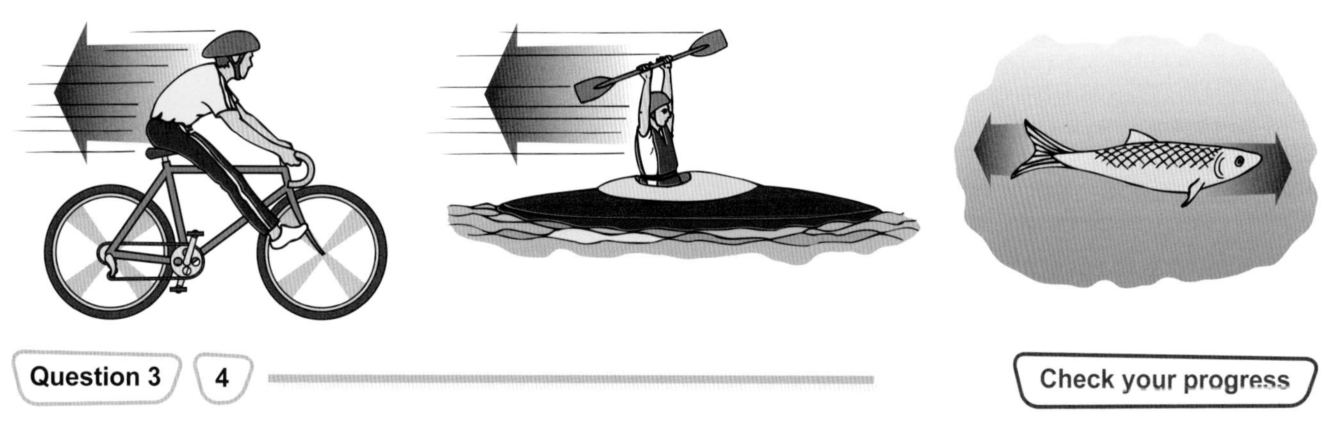

Question 3 **4**

Check your progress

9K.5 More about friction

You should already know | Outcomes | Keywords

Friction with the air

When an object moves through the air there is a friction force from the air that tries to slow the object down. We call this friction force **air resistance** or **drag**.

The size of the drag on a moving object depends on things like:

- the shape of the object;
- how smooth the surface of the object is;
- how fast the object is moving;

If you decrease the drag on something, it can go faster without increasing the driving force.

This idea is important in the design of cars. A shape that produces a low drag is **streamlined**. A streamlined car will use less fuel than a car without a streamlined shape on an identical journey.

The idea of streamlining is used in racing cycle design

- smooth bodysuit
- shaved legs
- specially shaped helmet
- cyclist bent low
- no mudguards.

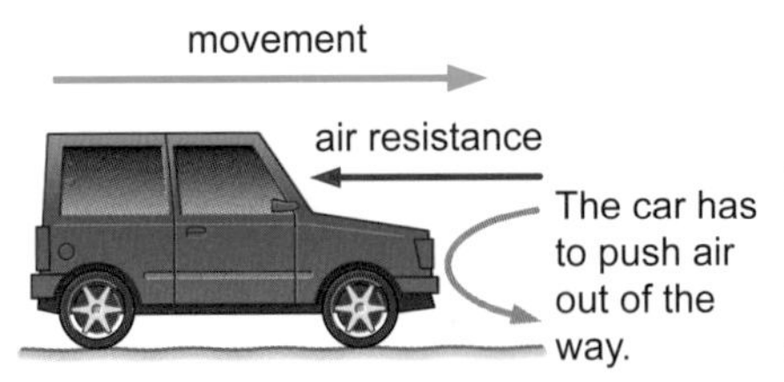

The shape of this van gives it a lot of resistance.

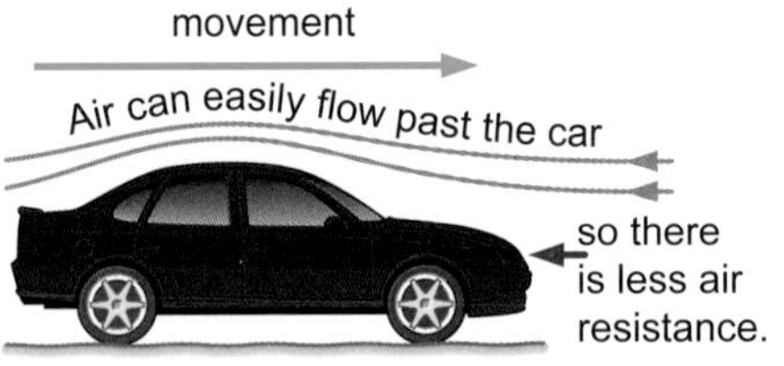

We say this car has a streamlined shape.

Drag gets bigger when something moves more quickly. This means that a car will use more fuel at a higher speed.

Diesel and petrol are manufactured from fossil fuels. We can reduce our consumption of these by designing streamlined cars, and getting drivers to slow down.

At 50 m.p.h. the car travels 50 miles on a gallon of petrol.

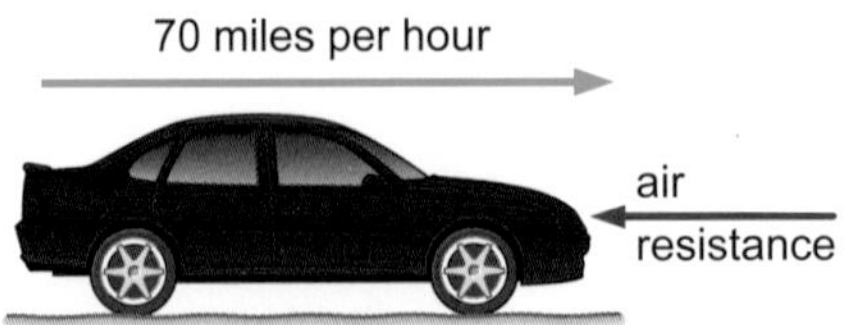

At 70 m.p.h. the car travels 30 miles on a gallon of petrol.

Car designers try to produce streamlined shapes that make the air flow smoothly around the car. This reduces the drag force.

Question 1 

Moving through water

The same idea applies when something moves through water. The friction force in water is often called drag.

The diagram shows the difference in the force needed to pull differently shaped blocks through water.

Fish have a streamlined shape that reduces drag. A fish uses less energy to swim than it would if its shape was not streamlined.

It takes a force of 3 N to pull this block of wood through the water.

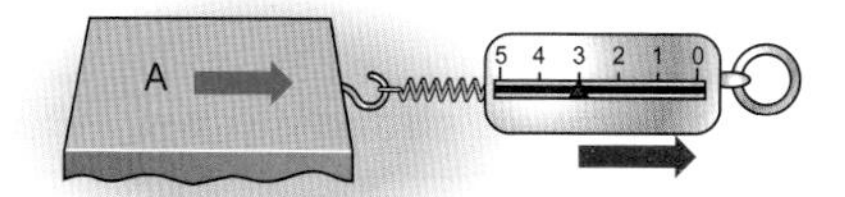

It takes a force of only 1 N to pull this block of wood through the water at the same speed.

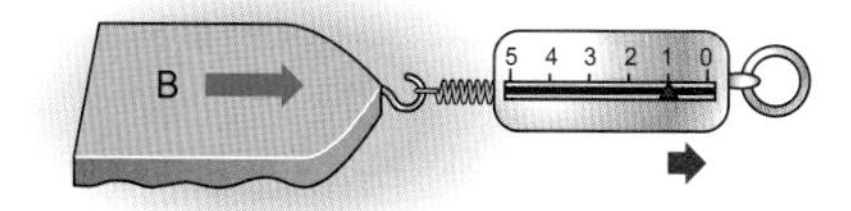

Block B has a more streamlined shape.

Question 4

Heating things

Friction warms things up. You can use friction to light a fire by rubbing wood together. This is sometimes done by making a stick of wood turn quickly at a point in the surface of a sheet of wood. A length of string on a bow is used to make the stick turn quickly.

You can use friction to light a fire.

When particles of dust from space enter the Earth's atmosphere, they are travelling at a high speed. The drag force makes them heat up so much that they burn. We see them as shooting stars.

The Leonid shower of shooting stars is seen every year around about November 17th when the Earth passes through a cloud of dust in space.

Leonid shower of shooting stars.

Drag also affects space capsules when they enter the atmosphere. The drag forces can increase the temperature of the capsule surface to 2000 °C. The surface must be made from a material that can stand this temperature. A heat shield is also needed to stop astronauts and sensitive equipment inside the capsule getting above 40 °C.

Question 5

Braking

We use a sudden increase in friction to slow down moving objects.

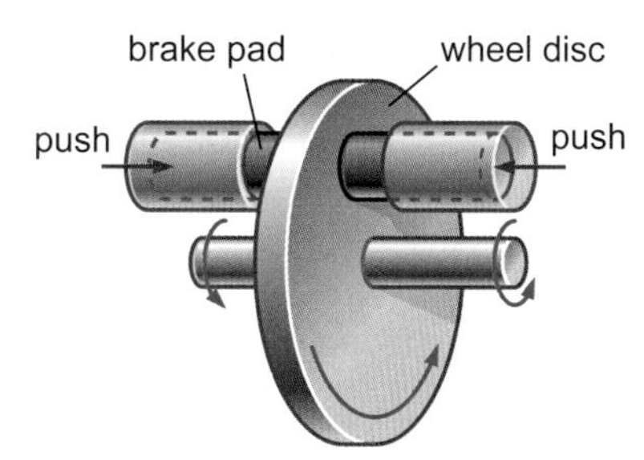

To slow a car down, the brake pads are pressed hard against the wheel disc.

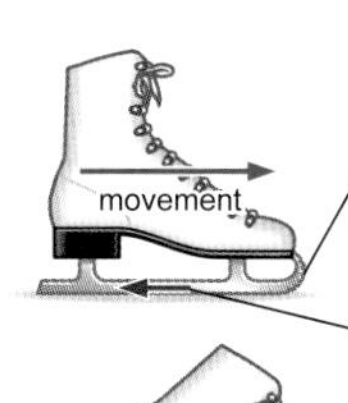

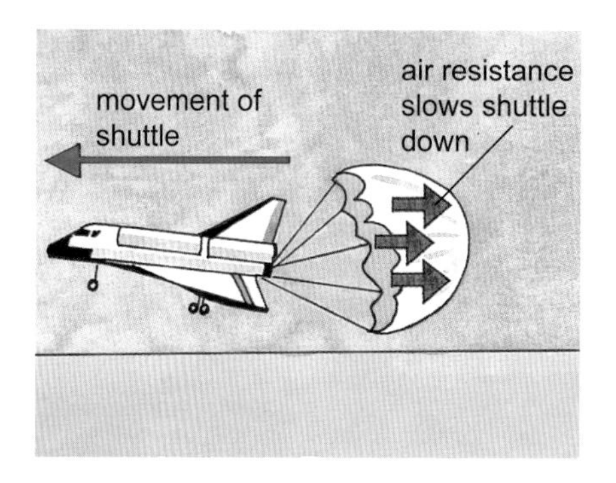

Question 6

9K.6 Skydiving

You should already know

Outcomes

Keywords

Terminal speed

When a skydiver jumps out of a plane, her weight makes her accelerate towards the Earth.

The skydiver's speed of fall increases by about 10 m/s every second at the moment when she leaves the plane because her weight is an unbalanced force. This is shown at point A on the graph and diagram below.

As the skydiver speeds up, air resistance starts to increase. Air resistance and the weight act in opposite directions so the total force accelerating the skydiver down is smaller.

The skydiver continues to speed up, but the speed increases at a slower rate. The acceleration is less. This is shown at point B.

After a little more time, the drag force has become even larger. It is only just less than the weight. The skydiver still accelerates but her speed is only increasing slightly. This is shown at point C.

Eventually the air resistance equals the weight and the skydiver does not accelerate any more. She falls at a steady speed called the **terminal speed**. This is shown at point D.

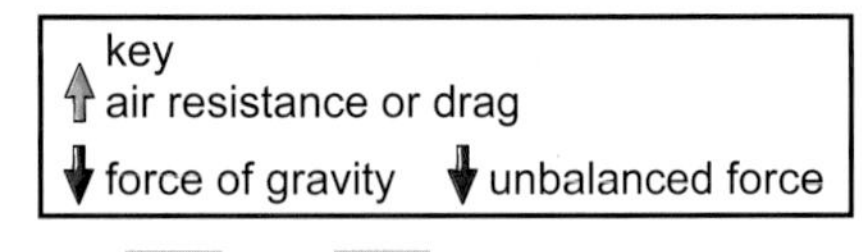

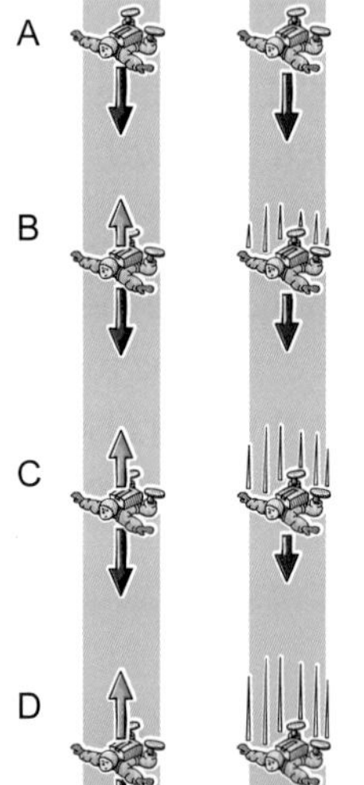

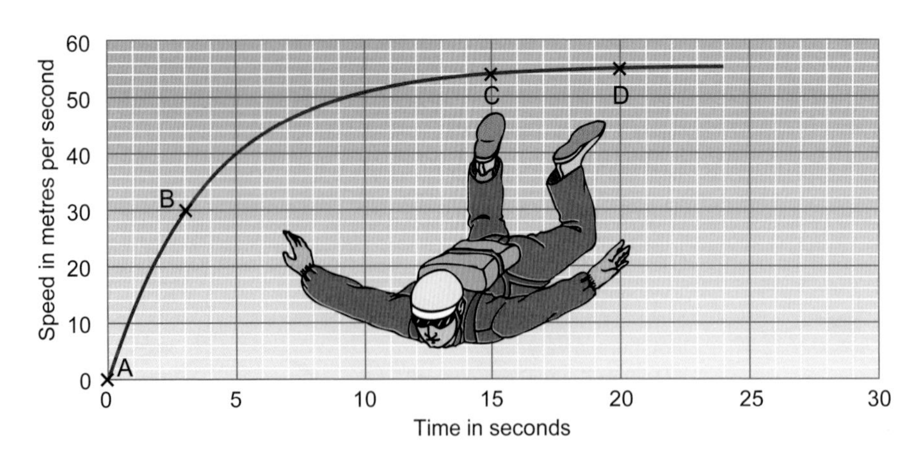

When the skydiver stops accelerating, we say that she has reached her terminal velocity.

Question 1 2

Using a parachute

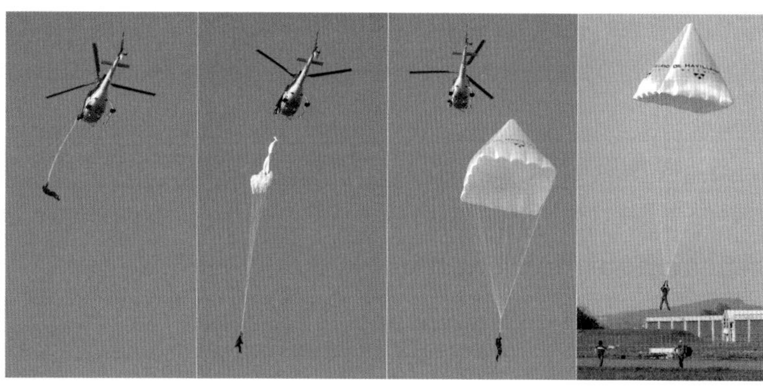

Testing Leonardo da Vinci's parachute.

When the skydiver reaches her terminal speed, she is falling at over 50 metres per second. If she hits the ground at that speed she will certainly be killed. To land safely, the skydiver must slow down to about 6 m/s. This is why she uses a **parachute**.

Parachutes were first mentioned in Chinese texts in about 90 BC. Leonardo da Vinci drew a design for a parachute in about AD 1480. It was not tested until a skydiver called Adrian Nicholas completed a 3000 m descent in a copy of Leonardo's parachute.

Opening a parachute suddenly increases the air resistance. This slows the skydiver down because the drag force is much bigger than the weight. As the skydiver slows down, the drag of the air on the parachute gets less. When the speed is about 6 m/s the drag equals the weight of the skydiver. The skydiver now falls at a new terminal speed of about 6 m/s. A skydiver who knows how to land can now land safely.

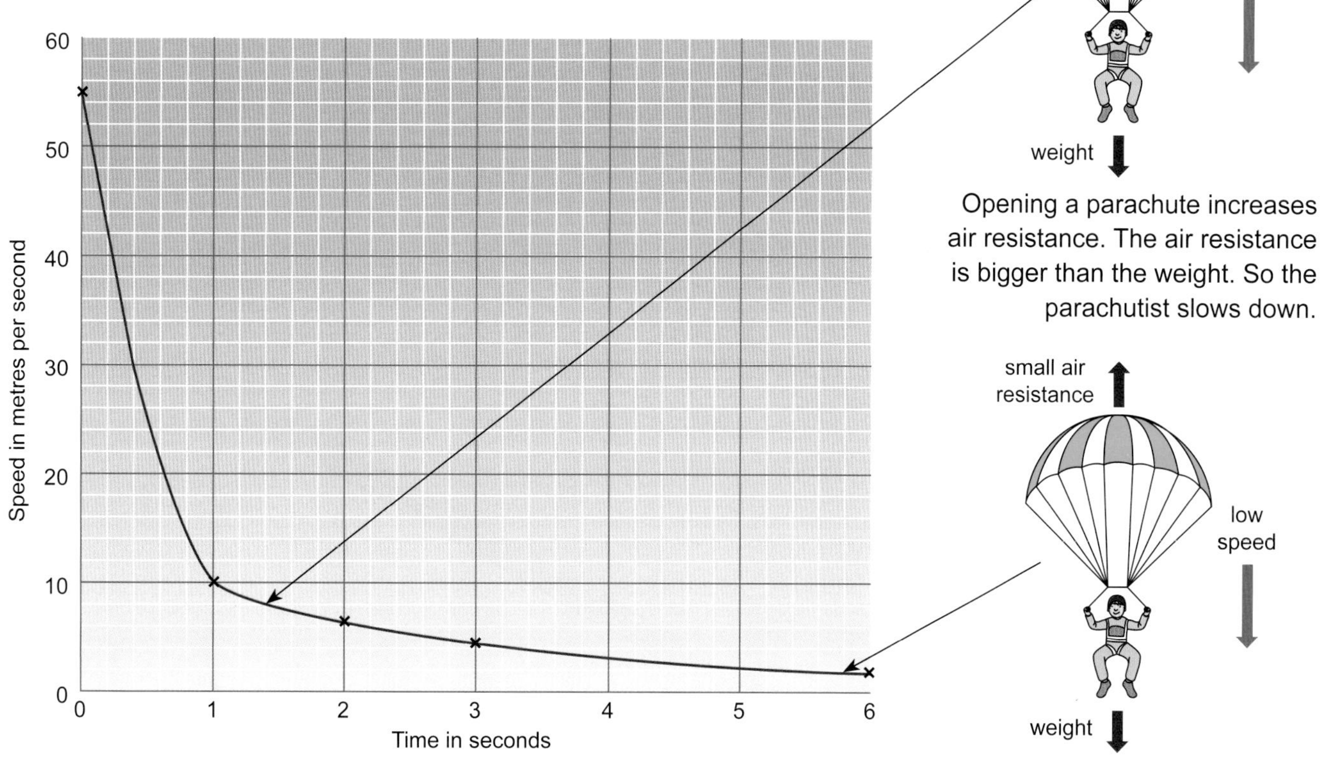

Opening a parachute increases air resistance. The air resistance is bigger than the weight. So the parachutist slows down.

Eventually the forces balance. The parachutist is falling at a lower speed.

Question 3 **4**

Review your work

Summary ➡

9K.HSW Reliable evidence

You should already know

Outcomes

Keywords

What do we mean by reliable evidence?

When scientists do experiments they collect **observations** and **measurements**. There is always a chance that someone making a measurement makes a mistake. One method used to eliminate errors in readings is to take an **average** of several readings. Doing this helps get a result that is **reliable**. If a result is reliable, it means that someone else can repeat your investigation and get the same answer.

Question 1 2

Valerie's experiment

Valerie did an experiment to find how the mass of a trolley affected the way it speeded up when you pulled it along. She decided to measure the speed of the trolley after it had been pulled over a distance from being released from still. She used a **light gate** connected to a computer to measure the speed after 30 centimetres.

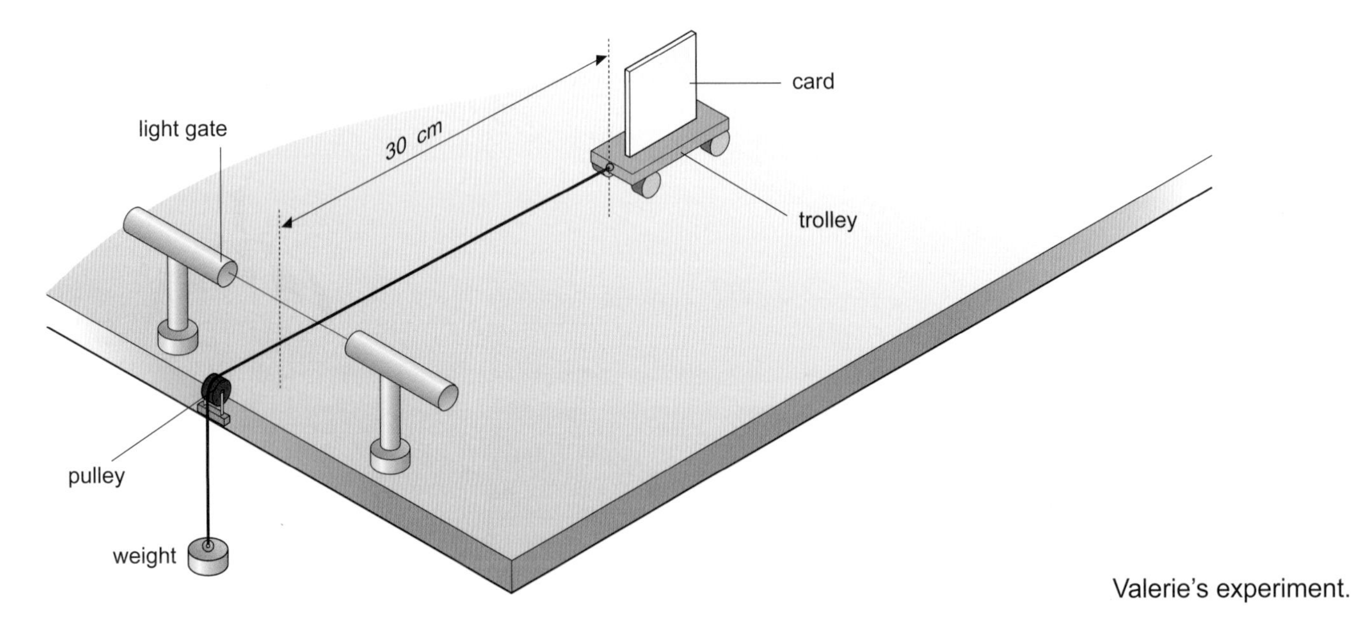

Valerie's experiment.

Valerie thought that the size of the pulling force from the weight would affect the speed of the trolley so she decided to use a weight of 5 N all the time. The force on the trolley is a **control variable**. It is kept the same, so it does not affect the outcome of the experiment. A second control variable is the distance from where the trolley is released to the light gate.

Valerie changed the mass of the trolley by putting weights on it. This was the **independent variable**. The independent variable is the thing you change. Some people like to think of it as the 'input' or the 'cause'.

The speed measurement is the 'output' or the 'effect'. This is called the **dependent variable**. The table shows Valerie's results.

Mass of trolley (kg)	Speed of trolley after 30 cm in m/s			
	Run 1	Run 2	Run 3	Average
0.5	2.49	2.47	2.43	2.46
1	1.73	1.74	1.76	1.74
1.5	1.43	1.89	1.41	1.58
2	1.22	1.21	1.24	1.22
2.5	1.15	1.1	1.09	1.11
3	1.1	1.01	0.99	1.03
3.5	0.93	0.94	0.99	0.95
4	0.61	0.87	0.87	0.78
4.5	0.81	0.89	0.81	0.84
5	0.77	0.79	0.75	0.77

There is a good chance of there being an error in every reading because Valerie is only human and humans make mistakes.

Valerie has repeated each reading three times and taken an average to avoid this. She is trying to make her result more reliable.

Taking an average helps to cancel out the errors if all your readings are scattered around the 'correct' value.

Sometimes a mistake gives a reading that is a very long way out. You should repeat it, or remember not to include it in your results.

A graph can be used to show results that might have a large error in them. They show up as a long way off any trend line. These are sometimes called **anomalous** results. You should check anomalous results to find out if they are mistakes.

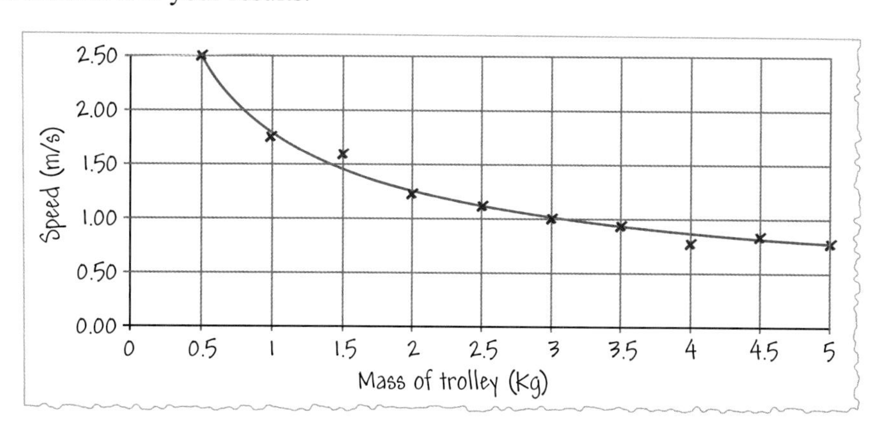

Valerie's results for speed against mass. There are two anomalous results off the line.

Question 3 

9K Questions

9K.1

1 What would suitable units be to measure the speed of:

a a sprinter?

b a high speed train?

2 a If a car travels at 70 km/h how many kilometres does it travel in two hours?

b How long would the car take to travel a distance of 210 km at 70 km/h?

3 a If a car is travelling at 120 km/h, is it breaking the British speed limit of 70 m.p.h.?

b Give a reason for your answer. If 5 miles is 8 km, how could you work out the British speed limit in km/h?

4 How does a speed camera or speed gun detect your speed?

5 a What is an anemometer used for?

b Suggest why it might be important to know the wind speed on a bridge over a river?

9K.2

1 Why is it only necessary to compare the times of runners in a race to compare how fast they have run?

2 a What was inaccurate about the timing methods used 50 years ago?

b How have methods improved?

3 a Work out the speed of the giraffe from the data in the diagram.

b How far does the giraffe travel in 10 seconds? Explain how you worked out your answer.

4 a Work out the speed of the shark from the data in the diagram.

b The shark and the swift are both very streamlined. Suggest a reason why the swift can fly faster than the shark can swim.

5 An eagle glides at 2 m/s.

a How far does it travel in 10 seconds?

b How long would it take an eagle to glide 100 m?

6 A skater slides on one skate across the ice at 1.5 m/s. How long does it take her to travel 9 m?

7 a If a car takes 4 hours to complete a journey of 160 miles, what is the average speed?

b Why is it important to refer to the speed calculated as the average speed for the journey and not just the speed?

9K.3

1 a What is the effect of friction on a moving object?

b Describe one way of reducing friction on a moving object.

2 a What are the light gates and computer used for in the experiment?

b Suggest one advantage of using this method of timing over using a handheld stopwatch.

3 What are the three forces acting on the glider and which of them is unbalanced?

4 What is meant by acceleration and what causes an object to accelerate?

5 What two things can an unbalanced force do to an object?

9K.4

1 Describe the effect the different forces have on the motion of the car as it travels down the slope and onto the sugar.

2 Give one practical application of a sudden increase in friction being used to slow something down.

3 a What happens to the friction force from the air as a car moves more quickly.

b Give two other terms for the friction force in different circumstances.

4 a What happens to the speed of something when the driving force is exactly balanced by the friction force?

b Give one practical example of this happening.

9K.5

1 a Write down three things that affect the size of the drag on a moving object.

b What is the effect on a moving object of decreasing the drag and keeping the driving force the same? Give a reason for your answer.

2 Why does a streamlined car travelling at the same speed as a non-streamlined van use less fuel?

3 a Write down one way that a car driver can reduce their fuel usage.

b Why is this important?

4 Why is it an advantage for a fish to have a streamlined shape?

5 Give two examples of friction heating something up.

6 Give three examples of a sudden increase in friction being used to slow something down.

9K.6

1 Why does the skydiver's speed stop increasing so quickly after the first few seconds of fall?

2 How does the parachutist's weight compare with the drag force at the terminal speed?

3 What is the effect of opening a parachute on a skydiver's terminal speed?

4 When it is first opened, a parachute produces a very large drag.

a What is the effect of this on the speed of the skydiver?

b Why does this drag force get smaller until the skydiver reaches terminal speed?

9K.HSW

1 Suggest some reasons why humans might make errors in taking readings in an experiment.

2 a What do scientists mean by a reliable result?

b Give one method of improving the reliability of a set of readings.

3 What does taking an average do to the errors in individual results?

4 What is an anomalous result and what should you do about it?

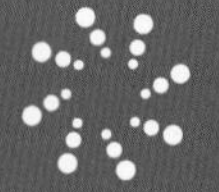

9L.1 What is pressure?

Scientific enquiry

You should already know

Outcomes

Keywords

The effect of a force

When a force pushes onto a surface its effect depends on two things:

- the size of the force;
- the size of the area it is acting on.

The soles of your feet have a small area. This is the area your weight acts on when you walk. Walking is not a problem on hard ground but you sink in if you walk on soft snow. This is because of the effect of your weight on the small area your feet are pressing on. The effect is called **pressure**.

Your feet sink into soft snow.

To stop yourself sinking into soft snow, you need to spread your weight over a bigger area than the soles of your feet. You can do this by wearing snow-shoes or skis. Spreading out your weight over a larger area gives a lower pressure.

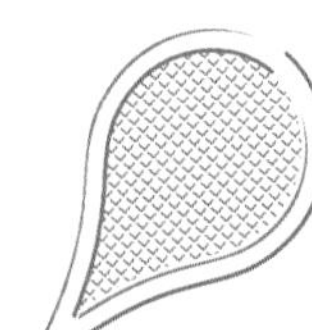

With an ordinary shoe you put all your weight onto a small area.

With snow shoes, the same weight is spread out over a much larger area.

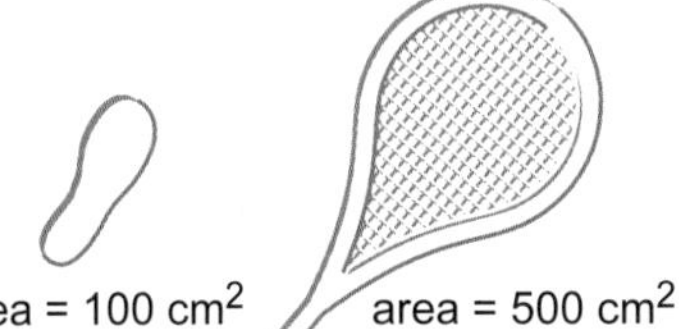

If you wear snow-shoes, your feet don't sink in.

The same idea applies to riding a bike. Thin tyres like those on a racing bike sink deeper into soft ground than wide tyres. The wider tyres like those on a mountain bike produce a lower pressure on the ground.

A racing bike sinks into the sand. It has thin tyres so there is a big pressure on the sand.

A mountain bike has fat tyres so there is a small pressure on the sand.

Question 1  

Changing the area to change the pressure

There are many situations in which we use a small area to get a big pressure and a large area to get a small pressure.

A drawing pin has a broad head and a sharp point.

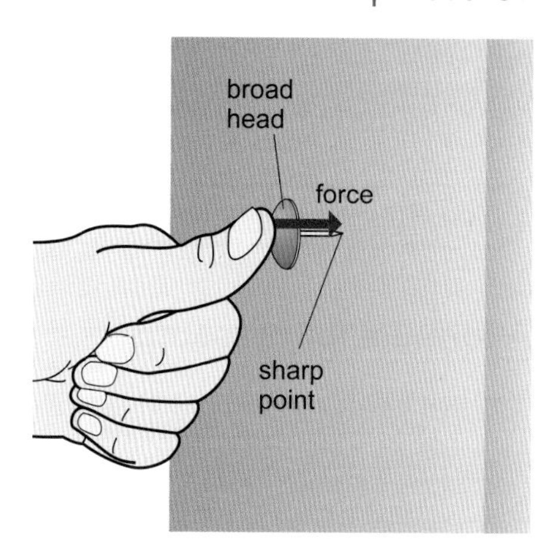

When you push it into a surface, the same force is pushing on your thumb as pushes on the surface. The area in contact with your thumb is large because the pin has a broad head. The pressure on your thumb is small so the effect of the force on your thumb is also small.

The area of the point of the pin is very, very small. This means the pressure on the point is very, very high. The effect of the force acting on the small area is to break into the surface that the pin is against.

In the drawing below, the truck tyres have a small area of contact with the ground. The truck has sunk into the mud. The tractor is being used to pull the truck out of the mud. The tractor tyres are very broad and have a large area of contact with the ground. It does not sink because the large area of contact produces a low pressure on the ground.

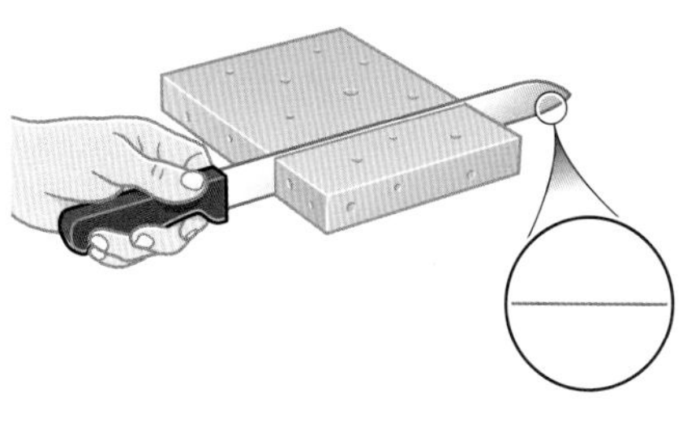

The sharp edge of a blade is very thin, so there is a big pressure on the cheese.

Question 4

More examples

The builder on the roof is kneeling on a long, wide board to spread her weight out.

This reduces the pressure on the roof. If the builder knelt directly on the roof the pressure would be too high. The builder would fall through the roof.

A sharp knife has a very thin edge so it creates a high pressure on anything you press it against, and it cuts easily.

A blunt knife can have an area on its cutting edge that is 100 times bigger than a sharp knife. The pressure it produces is too low to cut.

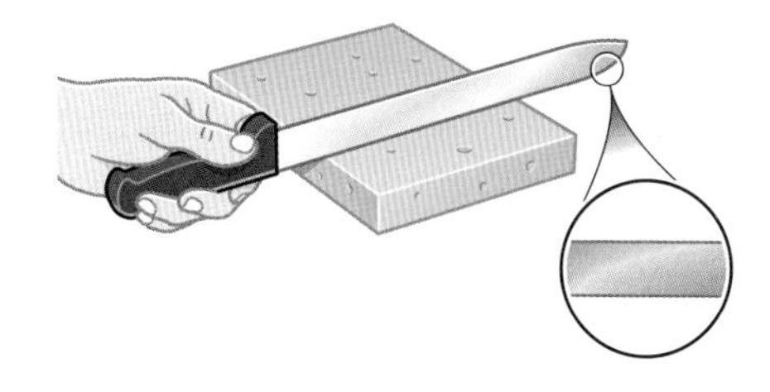

It's a lot harder to cut cheese with the blunt edge of a blade. This edge of the blade is 100 times thicker.

Question 5  6

9L.2 How big is the pressure?

You should already know | Outcomes | Keywords

A formula for pressure

You can calculate the pressure using a **formula**. The formula is:

pressure = force ÷ area

Force is measured in newtons. The area can be measured in different units depending on how big the area is. The **unit** of pressure depends on the units used for force and area.

In science, we usually measure:

- force in newtons (symbol N);
- area in square metres (symbol m^2).

The unit of pressure is the force unit divided by the area unit. This means the unit of pressure is **newtons per square metre**. The symbol for this is N/m^2.

Sometimes it is easier to use an area in cm^2 or in mm^2. If you do this, the pressure units change. The table shows you how.

Area unit	Pressure unit
cm^2	in N/cm^2
mm^2	in N/mm^2

The pressure tells you how much force acts on a unit of area. For example, if we press an area of $4\,mm^2$ with a force of 8 N, the pressure is given by:

pressure = force ÷ area

pressure = $8\,N \div 4\,mm^2$

pressure = $2\,N/mm^2$

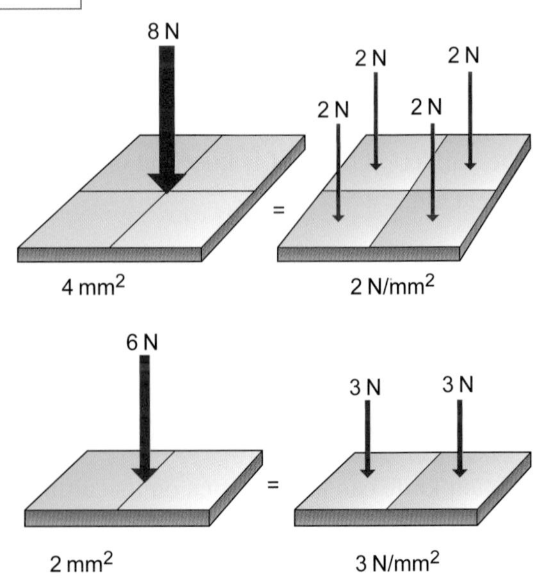

This means that there is a force of 2 N on each square millimetre.

The diagram shows this example and also what happens when a force of 6 N acts on an area of $2\,mm^2$.

Question 1  2  3

Working out the pressure on the ground

The gardener's wheelbarrow has a fat tyre. This means it has a large area of contact with the ground. The area of the tyre in contact with the ground is 80 cm². The force of the wheel on the ground is 400 N.

We work out the pressure by dividing the force by the area. This is how to show the working out:

pressure = force ÷ area

pressure = 400 N ÷ 80 cm²

pressure = 5 N/cm²

The fat tyre means that the weight is spread over a large area. So the tyre doesn't press very hard on the ground. We say that there is only a small pressure.

The gardener changes to a barrow with a thin wheel. The pressure is a lot higher. This is because the area of contact between the wheel and the ground is now only 20 cm². The wheel sinks into the ground.

We can work out the new pressure using 'pressure = force ÷ area'.

pressure = force ÷ area

pressure = 400 N ÷ 20 cm²

pressure = 20 N/cm²

The pressure on the ground for the thin tyre is four times as big as it is for the fat tyre.

Question 4 **5**

The thin wheel sinks into the ground because the pressure is high.

The pressure on you

The pull of the Earth on your body is called your weight. This is the force that presses on the area where you make contact with the surface you are on.

When you stand up, the area that your weight presses on is the soles of your feet. When you lie on a bed, your weight is pressing over a much larger area so the pressure is less.

If you can float on water, it is very relaxing because the pressure is even lower as the water shapes to match your body. This makes the area of contact as big as possible.

If the person in the diagrams has a weight of 600 N then the pressure on his feet when he is standing up is 12 000 N/m². The pressure on his body when he is floating on water is 1200 N/m². This is ten times lower.

Question 6

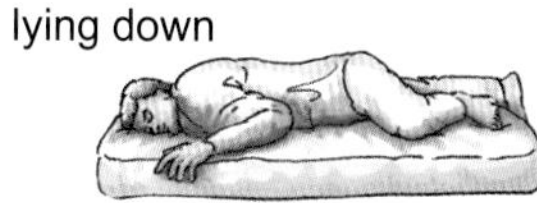

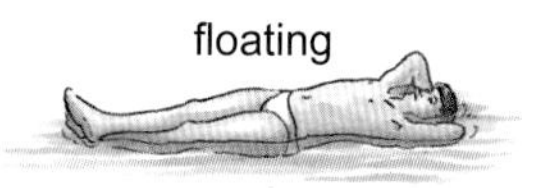

Check your progress

9L.3 Turning things

You should already know

Outcomes

Keywords

Making things turn

If something is free to move then a force can make it speed up, slow down or change direction.

Sometimes something is fixed so it cannot be moved along but it can move round.

A door is a good example. If you push on a door that is standing open, it does not move in a straight line because of the hinges. The combination of the force and the hinges makes the door turn. We say the force has a **turning effect**.

turning effect of force

force

The hinges are the point around which the door turns. This point is called the **pivot**.

The same effect happens when a child pushes a playground roundabout.

push

pivot

If you push like this the roundabout goes round the same way as the hands of a clock. We say it turns clockwise.

movement

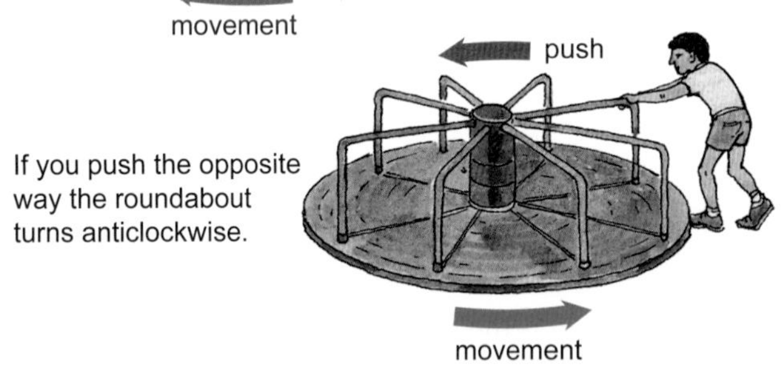

If you push the opposite way the roundabout turns anticlockwise.

Question 1 2

Moving a seesaw

When a child sits on the right-hand side of a seesaw and there is nothing to balance them on the other side, the seesaw turns in a clockwise direction.

The weight of the child has a turning effect about the pivot of the seesaw.

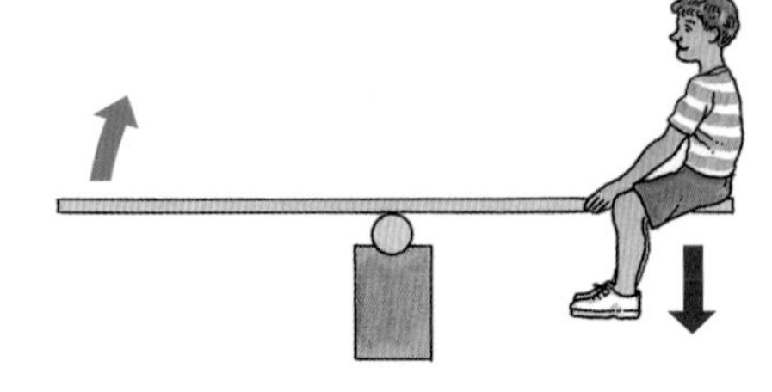

Question 3

Getting a bigger turning effect

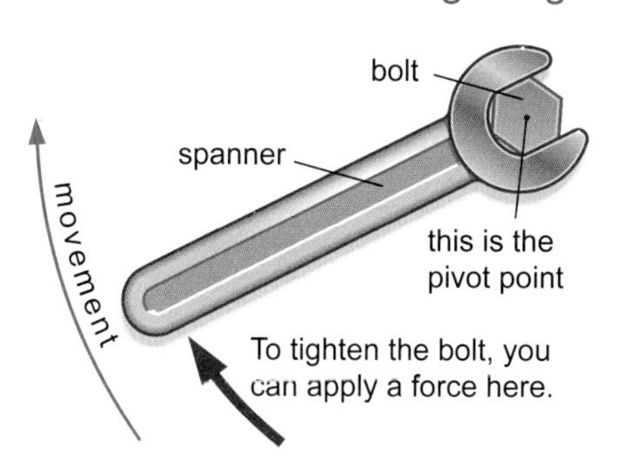

A spanner is used to tighten and slacken nuts and bolts.

You apply a force to the end of the spanner handle, and the centre of the bolt acts as the pivot.

Unless a nut or bolt is very slack, it is not possible to get a big enough turning effect by gripping it with your fingers alone. The handlc of the spanner gives you a bigger turning effect.

The size of a turning effect depends on the size of the force you apply and on the distance between the pivot and the place you apply the force to.

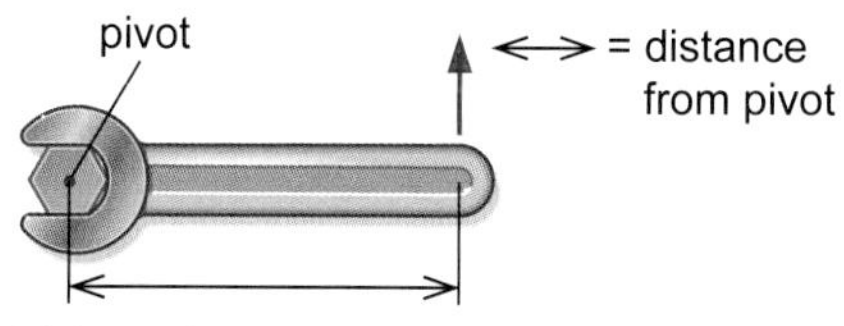

This nut is too tight to slacken with this spanner.

Sometimes a nut is too tight to undo with a normal spanner. This is very common with the nuts that hold car wheels on. They are put on in a garage with a power spanner and if you want to undo them to change a wheel by hand, they can sometimes be too tight.

One way to increase the turning effect you can get with a spanner is to put a metal pipe on the handle of the spanner to increase the distance between the force and the pivot.

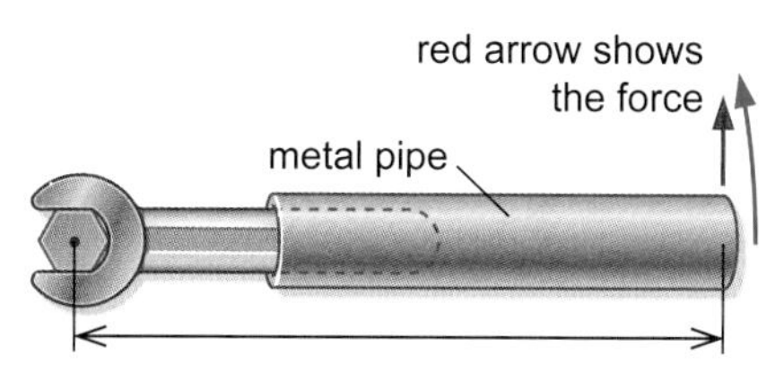

You can add a metal pipe to increase the distance between the force and the pivot.

Levers

The spanner is an example of a **lever**. We use levers to make things turn about a pivot.

Sometimes we apply the force a long way from the pivot to get a big turning effect. This effect is used with a spanner. The tool shown in the drawing uses the same idea to extract a nail.

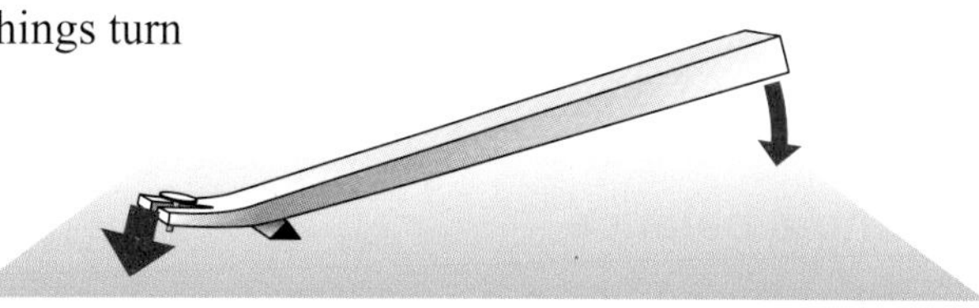

Extracting a nail.

Your arms and legs are examples of levers. Your joints are pivots. The muscles are attached to the bones. They apply a force that produces a turning effect around the pivot, and your arm or leg bends or straightens.

Muscles cannot push – they can only pull. This means you need two muscles for each lever in your body. One muscle produces a clockwise turning effect. The other muscle produces an anticlockwise turning effect.

In the diagram, the biceps muscle links the shoulder blade to a bone in your arm called the radius.

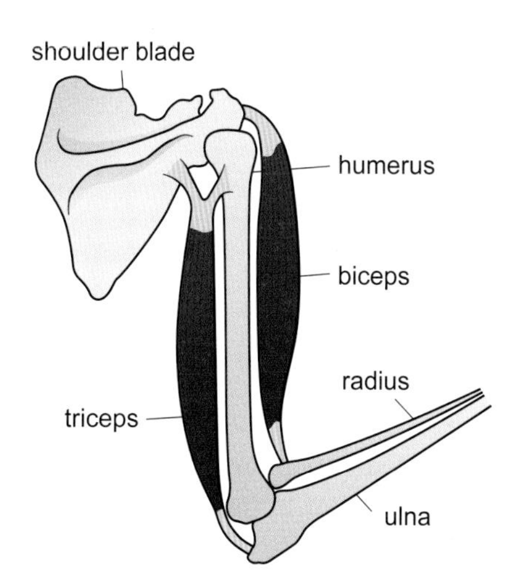

The elbow joint is the pivot. The biceps muscle pulls on the radius bone to move the forearm up. The triceps muscle pulls on the ulna bone to move the forearm down.

9L.4 Moments

You should already know | Outcomes | Keywords

What is the size of the turning effect?

The size of the turning effect produced by a force and a pivot is called the **moment**.

You can work out the moment of a force with this formula.

moment of force = size of force × distance from the pivot

Because a moment is calculated by multiplying a force in newtons (N) by a distance in metres (m), the units of the moment are Nm, which stands for **newton metre**.

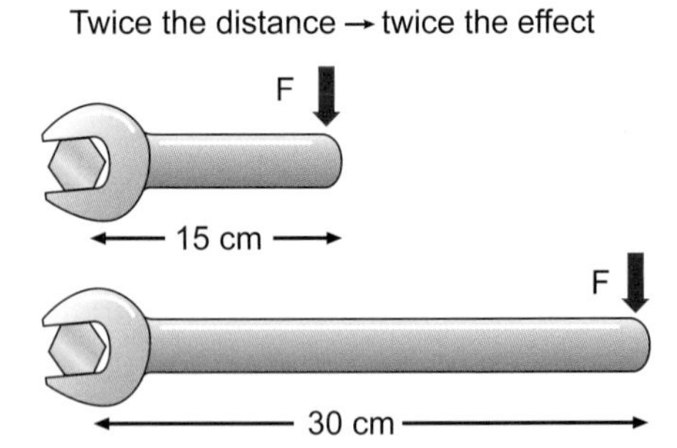

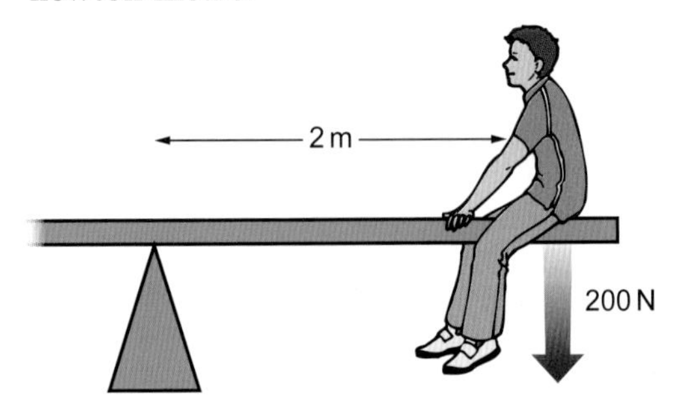

The moment of this boy on the seesaw
= force × distance of force from pivot
= 200 N × 2 m
= 400 Nm clockwise

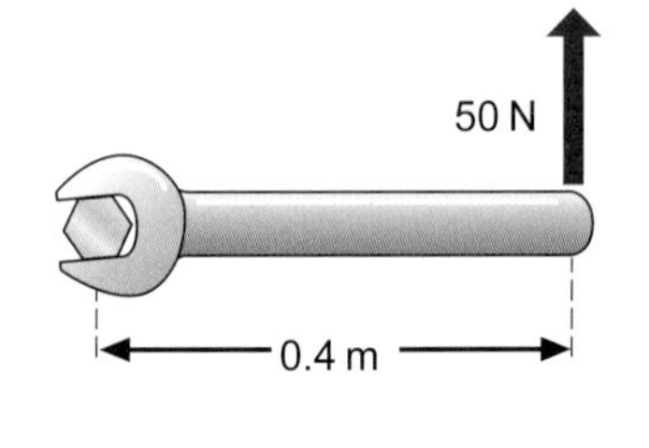

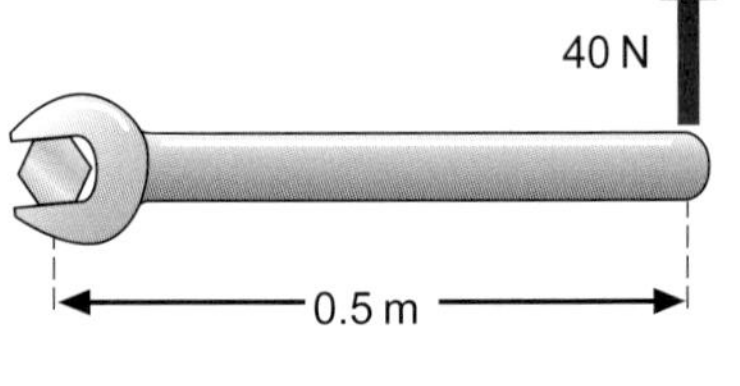

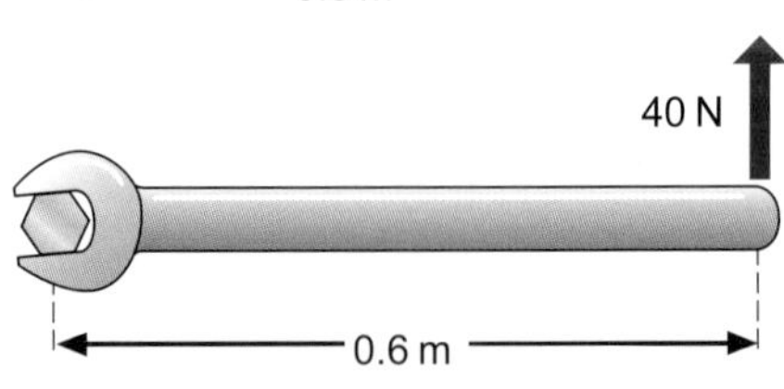

Question 1 | 2

Getting a balance

To make something balance, the turning effects of the forces acting on it must balance.

If the moment of the force turning a beam clockwise is the same as the moment of the force turning it anticlockwise then the beam will balance.

This is what happens on a seesaw.

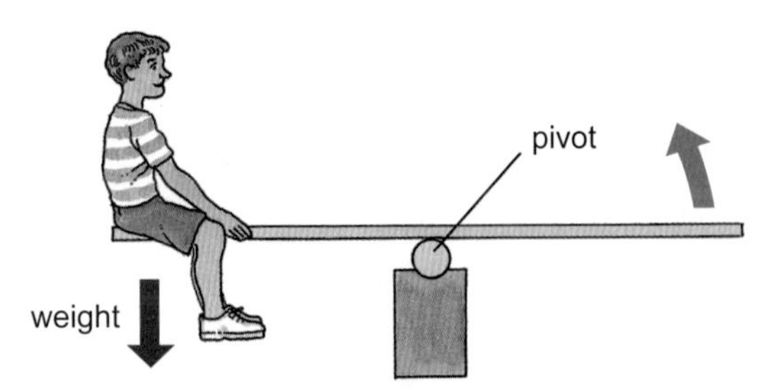

There is an unbalanced anticlockwise moment. So the seesaw turns anticlockwise.

Clockwise and anticlockwise moments balance. So the seesaw does not move.

Question 3

Balancing moments

You can balance a small object with a larger object if they both have the same moment. This is what happens when the small boy gets on the seesaw with his big sister. The table shows the moments.

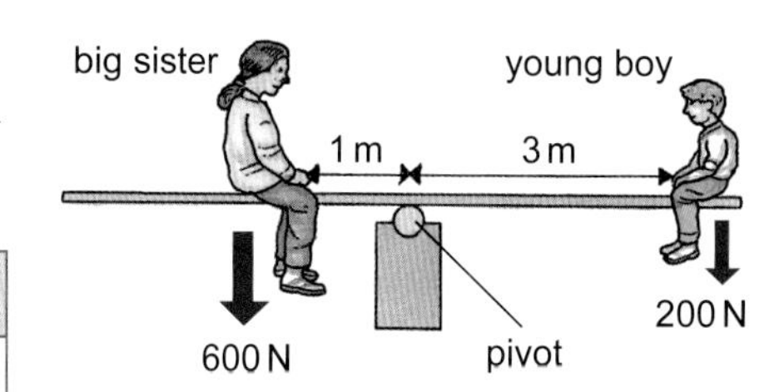

Who?	Force	Distance	Moment	Direction
young boy	200 N	3 m	600 Nm	clockwise
big sister	600 N	1 m	600 Nm	anticlockwise

The clockwise moment equals the anticlockwise moment, so the young boy and his big sister balance each other on the seesaw.

Using moments to weigh yourself

The diagram shows how a girl weighs herself using a plank and a 10 kg bucket of potatoes. She needs to remember that 1 kg has a weight of 10 N. That means that the weight of the bucket of potatoes is 100 N.

The girl weighs a lot more than 100 N. To balance the plank, she needs to stand quite close to the pivot. The potatoes need to be a good distance from the pivot.

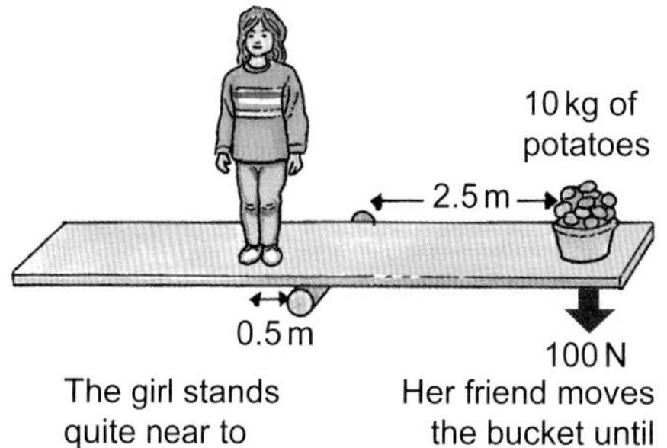

The girl stands quite near to the pivot.

Her friend moves the bucket until the plank balances.

The data is:

Weight of potatoes	100 N
Distance of potatoes from the pivot	2.5 m
Weight of girl	unknown
Distance of girl from the pivot	0.5 m

Because the beam is balanced, she can write this equation:

anticlockwise moment of girl = clockwise moment of potatoes

This means that:

weight of girl × distance of girl from pivot = weight of potatoes × distance of potatoes from pivot

Using the data table, the moment of the potatoes is:

$100\,\text{N} \times 2.5\,\text{m} = 250\,\text{Nm}$

The moment of the girl must also be 250 Nm when the beam balances.

This means the weight of the girl must be 500 N because:

$500\,\text{N} \times 0.5\,\text{m} = 250\,\text{Nm}$

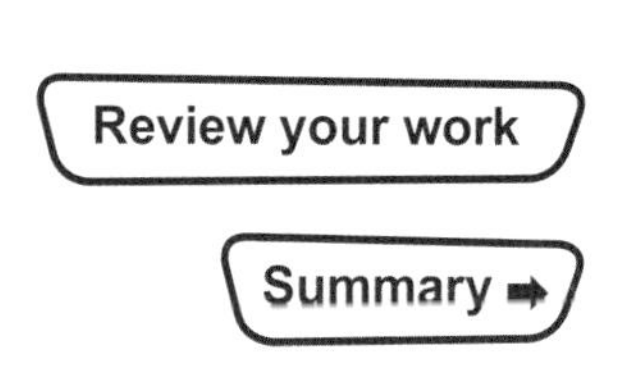

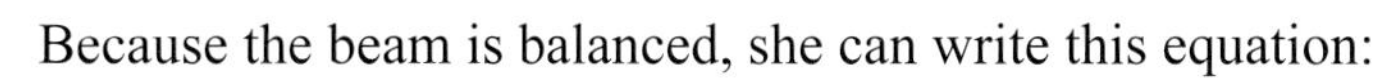

9L.HSW How secure is the conclusion?

You should already know | **Outcomes** | **Keywords**

How good is the data?

When scientists carry out experiments they collect data. The diagram shows the experiment to find the girl's weight you saw earlier.

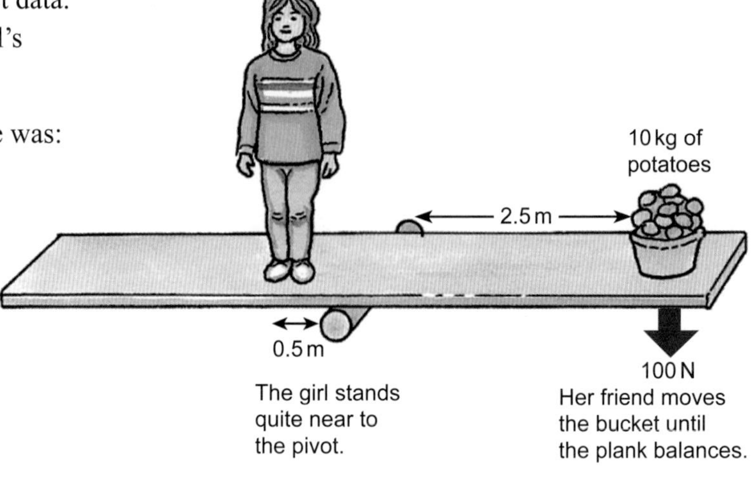

The girl stands quite near to the pivot.

The data used to work out the girl's weight before was:

- the distance from the girl to the pivot;
- the distance from the potatoes to the pivot;
- the weight of the potatoes.

The answer for the girl's weight was 500 N. But, how reliable is that answer? It depends on how good the three measurements are that are used to calculate it.

To get reliable measurements, the experimenter needs to make sure that:

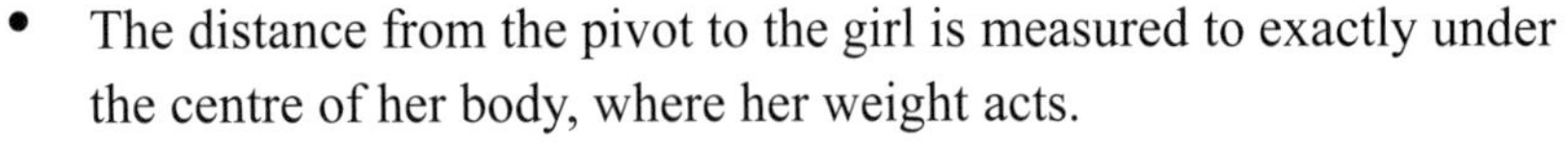

- The distance from the pivot to the girl is measured to exactly under the centre of her body, where her weight acts.
- The distance from the pivot to the potatoes is measured to exactly under the centre of the bucket of potatoes.
- They can work out where the centre of the bucket and the centre of the girl's body are.
- The balance used to find the mass of the bucket of potatoes gives an accurate answer.
- The plank balanced before the girl and the potatoes were in place. If it did not balance, it means the weight of the plank will affect the result.

The way an experiment is carried out is called its **implementation**. Good implementation will address the questions above so that the experiment produces **reliable evidence**.

Question 1  2  3

Errors in measurement

There are always errors in readings when you do experiments. This happens because of many things, including:

- the way the experimenter uses the equipment;
- the sensitivity of the equipment used;
- whether the equipment has been set up correctly.

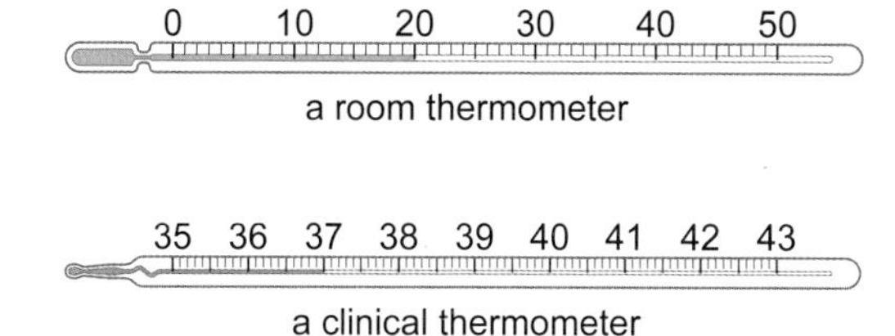

a room thermometer

a clinical thermometer

The clinical thermometer is more sensitive to small temperature changes.

One common error is called **parallax** error. The drawing shows what happens.

If you put the ruler on the object so there is a gap between the scale and the object, you can get a parallax error. The gap is caused by the thickness of the ruler. If you look at the scale at an angle, you get a reading that is different from the one you get if you look at it straight on.

To get round this, put the ruler edge on to the object you are measuring so the scale touches it. A clear plastic ruler with the scale on the underside can also be used. On this type of ruler the scale touches the object.

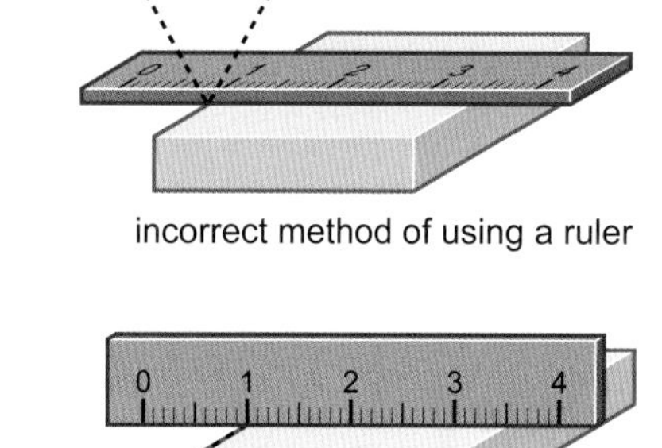

incorrect method of using a ruler

correct method

Question 4

More errors in implementation

Some errors are due to the equipment you use. They are called **systematic** errors.

Problem	Impact on results	Possible cure
clock used for timing is running slow	all timings recorded with lower times than they actually were	change clock for an accurate one or check it against an accurate clock and correct readings
end of a rule is damaged so it is 1 mm short	all readings are 1 mm short	add 1 mm to all readings

You combat these problems by checking your equipment carefully for accuracy and using it in the way it was intended by the manufacturer.

You may still get random errors in your readings because you are only human. You eliminate these by taking several readings and finding the average. Repeating readings and finding the average cancels out random errors. It makes your evidence more reliable. Reliable evidence leads to a more **secure conclusion**.

Question 5 **6**

9L Questions

9L.1

1 What does 'pressure' mean in terms of force and area?

2 a What two things does pressure depend upon?

b How can you reduce the size of the pressure acting on a surface?

3 Give two practical examples of pressure being deliberately reduced to make things easier.

4 Explain the design of a drawing pin in terms of pressure.

5 Why does a builder use a board to kneel on when working on a roof?

6 Why does a sharp knife cut better than a blunt one?

9L.2

1 What is the formula used to calculate pressure?

2 If a force is measured in newtons and an area is measured in square metres, what will the unit of pressure be?

3 A large packing case has an area of $2\,m^2$ and a weight of 200 N.
Calculate the pressure on the floor.
Remember to give the correct unit for the pressure.

4 Calculate the pressure on the ground for a racing cycle if the area in contact with the ground is $2\,cm^2$ and the weight of the cycle and cyclist is 800 N.

5 a The same cyclist rides a bike where the tyres have an area of $10\,cm^2$.
If the cyclist and bike weigh 800 N, what is the pressure this time?

b Why would the cyclist be less likely to sink into soft ground this time?

6 Why does lying down put less pressure on you than standing up?

9L.3

1. The combination of a force and a pivot can make something turn.
   What is the effect of the force called when this happens?
2. Give <u>one</u> everyday example of a force having a turning effect.
3. Why does the seesaw turn when a child sits on one of the ends and nobody sits at thc other end?
4. Give <u>two</u> ways of increasing the turning effect on something.
5. Describe <u>one</u> everyday application of increasing the turning effect on something.
6. What is a lever?
7. **a** Give <u>one</u> example of a lever in your body.

   b Describe where the pivot is and how the force or forces are applied.

9L.4

1. What is the name for the size of the turning effect of a force?
2. Work out the moment for each of the <u>three</u> spanners in the diagram.
3. A beam is balanced.
   What can you say about the moments acting on it?
4. What would happen if the young boy moved to a position 2 m from the pivot and his big sister did not move?
5. Where would his big sister have to move to in order to balance the beam if the young boy stays 2 m from the pivot?
6. What would the weight of the girl be if the plank balanced when she was 2.5 m from the pivot? (Assume the potatoes are still 2.5 m from the pivot.)

9L.HSW

1. **a** What are the <u>three</u> measurements needed to work out the girl's weight?

   b Why is it important that these measurements are reliable?
2. What is the word for the way an experiment is carried out?
3. Give <u>three</u> examples of things the experimenter needs to be sure of to get a reliable set of measurements in finding the weight of the girl.
4. What is a parallax error?
5. What is a systematic error and how do you overcome it?
6. What do you do to eliminate random errors?

9A Summary

Keywords

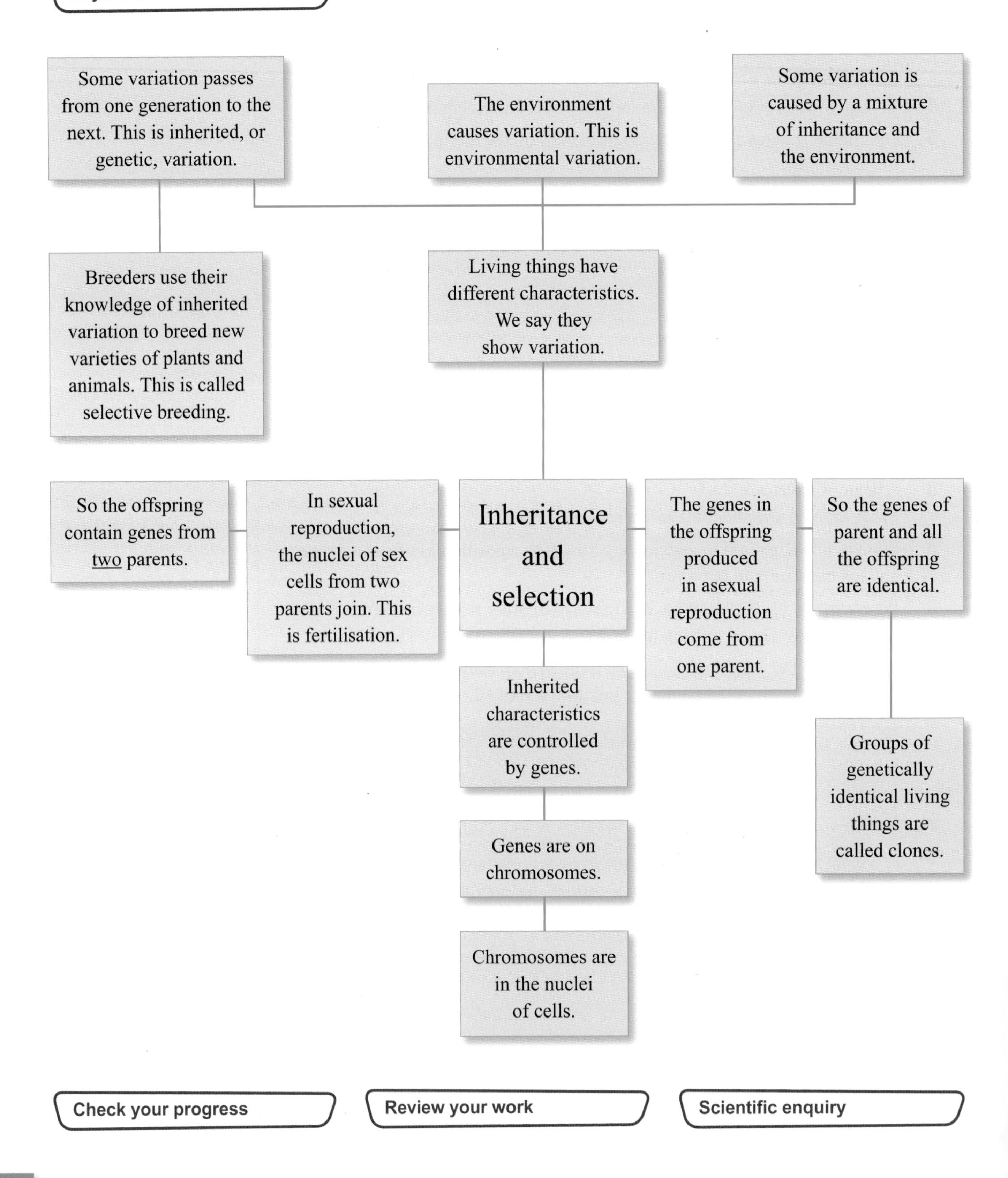

Check your progress

Review your work

Scientific enquiry

9B Summary

Keywords

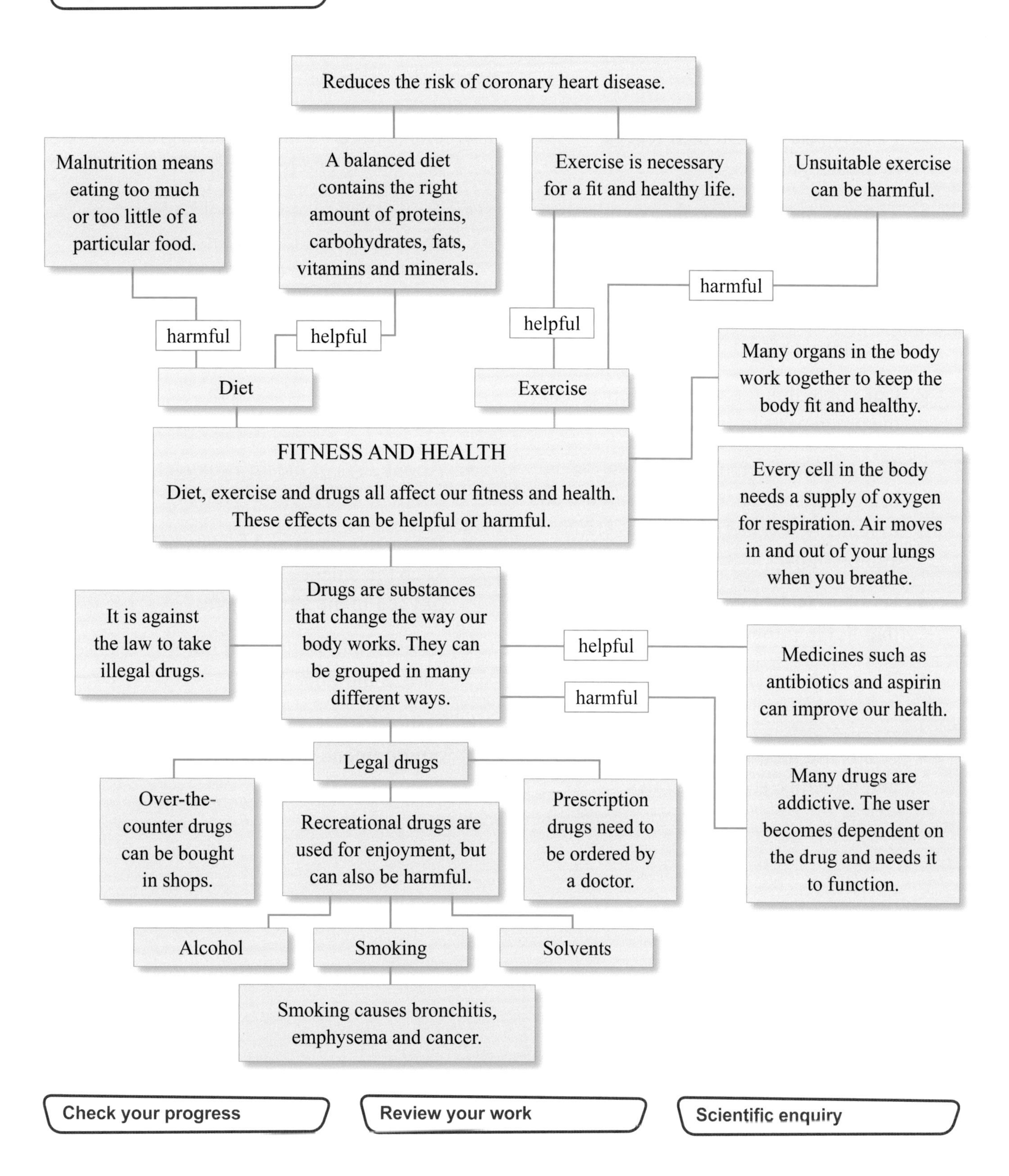

Check your progress

Review your work

Scientific enquiry

9C Summary

Keywords

Key ideas

- Plants make food in their leaves by photosynthesis.
- The raw materials for photosynthesis are carbon dioxide and water.
- The light energy needed for photosynthesis is trapped by chlorophyll.
- Chlorophyll is found in chloroplasts in the green parts of leaves.
- The glucose from photosynthesis is used to release energy in respiration and to make new materials such as starch and proteins.
- Biomass is the mass of the materials that living things are made of.
- Roots are branched and have root hair cells to increase their surface area.
- Plants use their roots to take up water and minerals, such as nitrates.
- Plants need nitrates to make proteins for making and repairing cells.
- A plant uses water to cool its leaves, to swell up fruits, to make its cells firm and to stretch its cells for growth.
- Most photosynthesis happens in the palisade cells near the top of a leaf.
- Plants produce oxygen in photosynthesis.
- About 21% of the air is oxygen and about 0.03% is carbon dioxide.
- Plants in oceans and forests help to keep the balance of carbon dioxide and oxygen in the atmosphere.
- Respiration, rotting and burning release carbon dioxide.
- Conservation is the protection of the environment, including living things.

Check your progress

Review your work

Scientific enquiry

9D Summary

Keywords

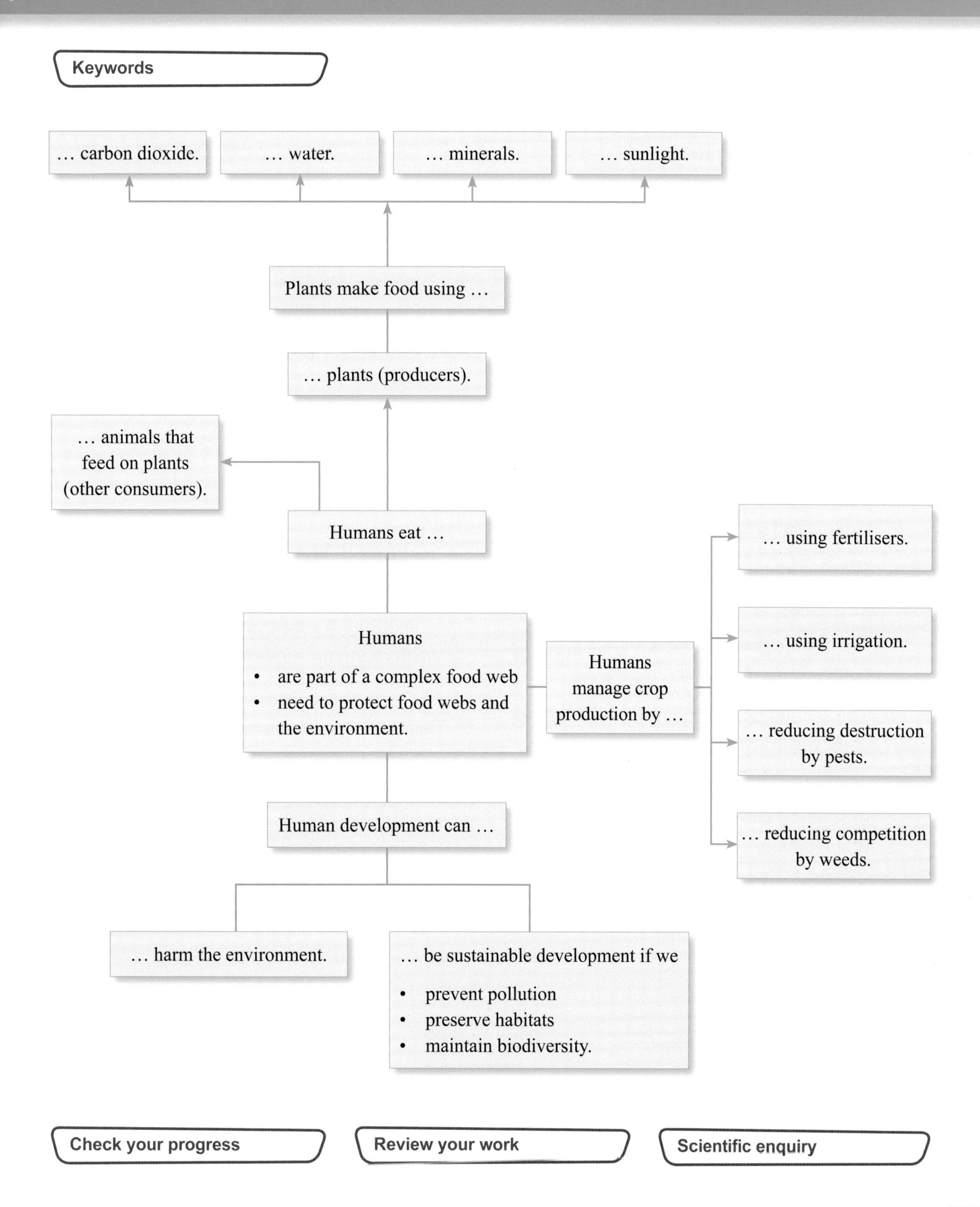

Check your progress

Review your work

Scientific enquiry

9E Summary

Keywords

Key ideas

- Metals are useful substances because of their properties.
- About three-quarters of the elements are metals.
- About one quarter of the elements are non-metals.
- Metals can be melted together to make a mixture called an alloy.
- Some metals will react with acid to make a salt and hydrogen.
- When a metal and an acid react, new substances are formed.
- The first part of the name of a salt comes from the metal.
- The second part of the name of a salt comes from the acid.
- A word equation shows the names of the substances at the start and the end of a reaction.
- A metal carbonate is a compound made from a metal with carbon and oxygen.
- A metal carbonate reacts with an acid to make a salt, water and carbon dioxide.
- A metal oxide is a compound made from a metal and oxygen.
- Most metals are extracted from substances we dig out of the ground, called metal ores.
- Most metal ores contain a lot of metal oxide.
- A metal oxide will react with an acid to make a salt and water.
- A combustion reaction is one where something burns and combines with oxygen.
- A thermal decomposition reaction happens when something is heated so it splits up into new substances.
- A reaction where an acid or an alkali react with something to form a neutral solution is called a neutralisation reaction.

Check your progress

Review your work

Scientific enquiry

9F Summary

Keywords

Key ideas

- Some metals like gold and silver do not react with water.
- Some metals like magnesium have a very slow reaction with water.
- Metals in Group 1 of the periodic table are very reactive when put into water.
- Metals in Group 1 of the periodic table are called alkali metals.
- If metals are put in the order of how well they react with water you get a reactivity series.
- Gold, silver and copper can be found as pure metal in the ground. These are called native metals.
- Most metals are extracted from rocks containing metal compounds. These rocks are called ores.
- A more reactive metal can take the place of a less reactive metal in a compound. This is called a displacement reaction.
- Carbon will displace some metals from their ores.
- Iron can be displaced from iron ore using carbon.
- Metals that are more reactive than carbon can be extracted from their ores using electricity.
- Alloys are mixtures of metals.
- Alloys have different properties from the metals they are made from.
- When a metal reacts with hydrochloric acid, the second part of the name of the salt is 'chloride'.

Check your progress

Review your work

Scientific enquiry

Keywords

Key ideas

- Soil contains plant and animal material called humus.
- Humus adds mineral nutrients to the soil when it breaks down.
- Compost is made from rotting vegetable and plant material. It is often added to soil by gardeners to improve it.
- Fertilisers can be added to the soil to add nutrients and minerals that plants need.
- An acid soil has a pH below 7.
- An alkali soil has a pH above 7.
- Rainwater is very slightly acidic.
- Pollution increases the acidity of rainwater.
- Acid rain has a negative effect on buildings and on living things.
- When the living things in an area gradually die out, it is called progressive depletion.
- Carbon dioxide gas contributes to acid rain.
- Sulfur dioxide gas contributes to acid rain.
- Nitrogen oxides released into the air by burning fuels contribute to acid rain.
- The gradual rise in the temperature of the Earth's surface is called global warming.
- Scientists think that gases like carbon dioxide in the atmosphere are making it act like a greenhouse, keeping in the heat from the Sun. This is called the greenhouse effect.
- The greenhouse effect is causing global warming.

Check your progress

Review your work

Scientific enquiry

9H Summary

Keywords

Key ideas

- In a combustion reaction something usually combines with the oxygen in the air when it burns.
- A fuel is a substance which we burn to get heat.
- If something burns and there is not enough oxygen for it to produce carbon dioxide gas, it may produce carbon monoxide gas. This is called incomplete combustion.
- Oxidising agents are substances that contain oxygen. They are often used in fireworks and explosives.
- A cell is a combination of two electrodes and a third substance. It is used to make electricity.
- A chemical reaction that produces heat is called an exothermic reaction.
- Biochemistry is the study of the substances in living things.
- Petrochemistry is the study of substances, fuels and plastics made from oil.
- Polymer chemistry is the study of plastics and fibres.
- The pharmaceutical industry makes drugs and medicines.
- Agrochemistry is the study and manufacture of substances used in farming.
- When you make a new substance out of some others, it is called synthesis.
- A synthetic material is one that is made from other materials.
- Sustainable development means improving the living conditions for humans without having a negative effect on the Earth's resources.
- A join between atoms is called a bond.
- You have to put in energy to break bonds. Energy is released when bonds form.
- An exothermic reaction gives out heat because the bonds that form give out more energy than was needed to break the bonds at the start of the reaction.
- The mass of the products of a reaction is the same as the mass of the substances present at the start. This is called the conservation of mass.

Check your progress

Review your work

Scientific enquiry

9I Summary

Keywords

Key ideas

- Whenever anything happens, energy is changed from one type into another or moves from one place to another.
- Chemical energy is stored in fuels like oil, coal and gas. It is also stored in food and in cells and batteries.
- Photosynthesis is the process in green plants where energy from sunlight is changed into chemical energy in food.
- When an object is lifted to a higher position above the Earth's surface, its gravitational potential energy is increased.
- The voltage of a cell or battery is a measure of how much energy it transfers to a circuit when a current of 1 A flows for one second.
- The different parts of a circuit are called components.
- An appliance is something that runs on electricity, like a hairdryer or a toaster.
- The power of an appliance is the number of joules of electrical energy it changes into other types of energy every second.
- A change of energy of one joule per second is called 1 watt.
- A kilowatt is 1000 W.
- 1 watt is written as 1 W.
- 1 kilowatt is written as 1 kW.
- When anything happens, the total amount of energy at the end of the process is equal to the total amount of energy at the start. This is called the conservation of energy.
- Electricity can be generated by moving a coil of wire near a magnet. This is done on a large scale in a generator.
- A small generator is often called a dynamo.
- In a power station a turbine is driven round using steam and this is used to drive a generator.
- Many power stations use fossil fuels to heat water to produce steam. The use of fossil fuels produces carbon dioxide which contributes to global warming.

Check your progress

Review your work

Scientific enquiry

9J Summary

Keywords

Key ideas

- Gravity is a force that attracts objects together.
- The size of the force of gravity gets bigger if either of the masses involved gets bigger.
- The size of the force of gravity gets smaller if the masses involved get further apart.
- The pull of the Earth or a planet or a moon on an object due to the force of gravity is called weight.
- Weight is a force, so its unit is the newton (N).
- The pull of gravity on the Earth's surface is 10 N for every kilogram. This is called the gravitational field strength.
- There are eight planets in the Solar System including the Earth.
- Pluto used to be classed as a planet but now it is classed as a dwarf planet because it is a lot smaller than the rest of the planets.
- An asteroid is a large lump of rock in space. There are many asteroids going around the Sun in orbits like the eight planets.
- A comet is a ball of rock and ice which orbits the Sun.
- Planets, asteroids and comets orbit the Sun on paths called ellipses.
- The path for a comet is a very stretched ellipse so it only passes near to the Sun once every hundred years or so.
- A satellite is an object orbiting a planet.
- The Moon is a natural satellite of the Earth.
- An artificial satellite is a man-made object which is made to orbit a planet.
- Geostationary satellites orbit the Earth at the same rate as the Earth turns, so they stay over the same point on the Earth's surface.

Check your progress

Review your work

Scientific enquiry

9K Summary

Keywords

Key ideas

- Speed is how fast something moves.
- Units used for speed include m/s, km/h and m.p.h.
- An anemometer is used to measure wind speed.
- The average speed is the total distance travelled divided by the total time taken.
- Friction is a force that acts to prevent things moving past each other.
- A light gate is a sensor which can be connected to a computer to measure how long it takes an object to pass through the gate.
- An unbalanced force acting on an object that is free to move will speed the object up.
- An unbalanced force acting in the opposite direction to motion will slow an object down.
- Acceleration means speeding up or changing direction.
- When something moves through air, there is a force of friction called air resistance or drag.
- Objects like cars and aeroplanes are shaped to reduce the drag when they move through air. This is called streamlining.
- Objects that move through water, like boats and fish, feel drag from the water.
- Objects that move through water usually have a streamlined shape to reduce the size of the drag.
- A parachute produces a lot of drag so it will slow down an object falling through the air.
- When the drag on a falling object equals its weight, the forces are balanced and it falls at a steady speed called its terminal speed.

Check your progress

Review your work

Scientific enquiry

9L Summary

Keywords

Key ideas

- The effect of a force acting over an area is called pressure.
- The formula used to calculate pressure is:
  pressure = force ÷ area.
- The normal unit used for pressure is newtons per square metre (N/m^2).
- When a force acts at a distance from a pivot, it has a turning effect.
- The turning effect of a force is used in a lever.
- The size of a turning effect is called the moment.
- The moment is calculated by multiplying the force by its distance from the pivot.
- The unit for a moment is the newton metre (Nm).

Check your progress

Review your work

Scientific enquiry

Glossary/Index

carbon dioxide a gas in the air produced by living things in respiration, in combustion or burning, and when an acid reacts with a carbonate; used by green plants for photosynthesis 42, 100

carbon monoxide a poisonous gas 24, 108

cell uses a chemical reaction to push an electric current around an electric circuit 110

characteristics the special features of a plant or animal 2

chemical energy energy stored in fuels and food 120

chemical reaction a change where one or more substances changes into a different substance or substances by breaking and forming bonds between atoms 74

chloride second part of the name of a salt containing chlorine; reactions between metals and hydrochloric acid produce chloride salts 86

chlorophyll the green substance in chloroplasts that traps light energy 44, 58

chloroplast where photosynthesis happens in plant cells; the parts that contain chlorophyll 44

chromosomes structures containing genes and found in the nucleus of a cell 4

circulatory system group of organs involved in circulating blood around the body 20

climate change changes in climate patterns such as those caused by global warming 102

clones a group of genetically identical living things 12

combustion when substances react with oxygen and release energy; another word for burning 108

combustion reaction the reaction between a fuel and oxygen releasing energy 78

communicate pass on information, usually through speaking, writing or the use of charts and diagrams, email or visual presentations 128

comet a small body that travels around the Sun in an eccentric orbit; a comet is usually made of ice and dust, and shows a tail when it passes near to the Sun 134

compete when several plants or animals are trying to get the same things 60

components devices used in electric circuits 122

compost rotted down vegetable matter used to fertilize soil 96

conservation preserving or taking care of living things and their habitats 48

conservation of energy the idea that the total amount of energy in a system stays the same; energy is neither created nor destroyed, but it can change from one form into another 124

conservation of mass a scientific law stating that no loss of mass happens in a chemical reaction 114

consumers animals that cannot make their own food but instead eat plants and other animals 56

control part of an investigation that is needed to make a test fair; needed so that we can be sure of the cause of a change or a difference 42

control variables variables whose values are kept the same to make the experiment a fair test 92, 154

coronary heart disease disorder of the coronary arteries that take blood to the heart muscles 30

crop yields how much food or other useful material a plant or animal crop can produce 50

D

deficiency diseases disorders caused by lack of a particular nutrient in the diet 26

dependent variable the output from an experiment; this is the variable you measure as a result of what you do in the experiment 154

dephlogisticated air the name used for the gas we know as oxygen, from when scientists believed in the phlogiston theory of burning 116

digestive system all the organs that are used to digest food 20

direction the line along which something moves or points 146

displacement reaction when a more reactive element pushes a less reactive element out of one of its compounds 88

drag the force of friction on an object moving through a fluid like air or water 150

drug a substance that can change the way that your body works or to treat a disease 28

dynamo an electrical generator 126

E

ecosystem a system that includes all the interactions between the environment, plants, animals and micro-organisms in a place 66

electrodes the pieces of material that conduct electricity and from the positive and negative terminals of a cell 110

ellipse a circle that has been slightly squashed; if you cut a slice through a cone parallel to the base of the cone, you get a circular edge on the cut surface but if you cut the cone at an angle rather than parallel to the base, the edge of the cut surface is an ellipse 134

emphysema a condition caused by constant coughing that break the walls of the air sacs so the person struggles to get enough oxygen 22

energy energy is needed to make things happen 114, 120

environmental conditions conditions such as light and temperature in the environment 64

environmental variation differences within a species caused by the environment 2

ethics a set of principles and values that might show how to behave in a particular situation 14, 50

ethology study of the ways animals behave in natural surroundings; that is, not under experimental conditions 32

exothermic if something is exothermic, it gives out heat into its surroundings 114

exothermic reaction a reaction that gives out heat 110

F

fair test a test in which one variable is varied and other variables are controlled or kept the same 92

fats part of our food that we use for energy 26

fertilisation when a male sex cell nucleus joins with a female sex cell nucleus to start a new plant or animal 4, 10

fertilisers chemicals added to soil to provide the minerals that plants need to grow 58, 64, 96

fibre undigestible cellulose in our food; it prevents constipation 26

fit able to do a lot of exercise without tiring 24

fitness ability to do exercise without tiring 20

formula uses symbols to show how many atoms of elements are joined together to form a molecule of an element or a compound 144, 160

fossil fuel fuel formed in the Earth's crust from the remains of living things; for example, coal 126

friction a force when two surfaces rub past each other; it acts in the opposite direction to the direction in which something is moving 146, 148

fuels a substance that burns to release energy 108

G

generate make 126

generator produces electricity when it is supplied with kinetic energy 126

genes parts of chromosomes that control the inherited characteristics of plants and animals 4, 8

genetic engineering transferring genes from the cells of one organism into the cells of a different organism 14, 50

genetic information instructions for particular characteristics coded on the genes 4

geostationary orbit an orbit in which a satellite stays above the same point on the Earth's surface, all the time 136

global warming increase in the average temperature on Earth 48, 102, 126

glucose a carbohydrate that has a small, soluble molecule 44

gravitational field strength the pull of gravity on a one kilogram mass; at the Earth's surface, the pull of gravity on a 1 kg mass is 10 N so we say that the gravitational field strength at the Earth's surface is 10 N/kg 132

gravitational potential energy energy stored in objects high up 120

gravity the attraction of bodies towards each other 132

greenhouse effect the warming effect on the Earth of greenhouse gases in the atmosphere; without it, the Earth would be much colder 102

greenhouse gases gases such as methane and carbon dioxide that stop some of the heat escaping from the atmosphere 48, 102

H

habitat the place where a plant or animal lives 66

heart disease any disease in which the heart doesn't work properly; often used to mean coronary heart disease in which one or more arteries to the heart muscles is blocked 34

herbicide a chemical that kills plants 62

humus the decaying organic material in soil 96

I

immune system the system in your body that protects you from infection 34

implementation to do something, to put into effect 166

incomplete combustion when a substance doesn't burn completely because of lack of oxygen 108

S

salt a compound produced when an acid reacts with a metal or an alkali 74, 86

samples small parts, for example of a population, to get an idea of the whole 32

Sankey diagrams diagrams used to show the energy changes that occur during a process or in an appliance 128

satellite a body orbiting a planet 136

scientific journals specialist journals in which scientists report the results of their investigations 34

secure conclusion a secure conclusion is one which can be relied on 166

selective breeding breeding only from the plants or animals which have the characteristics that we want; also called artificial selection 6, 10, 14, 50, 64

sexual reproduction reproduction in which the nuclei of two sex cells join to start a new life 12

side effects bad effects of a useful drug such as headaches and nausea. 28

speed distance travelled in a certain time 142

starch a carbohydrate with large, insoluble molecules 44

streamlined designed to keep drag to a minimum 150

subjective something that is subjective depends upon the observer for example the perception of the colour of something 80

sulfur dioxide an oxide of sulfur that helps to produce acid rain 100

sustainable systems that are managed so they can keep on going 48

sustainable development development in a way that won't damage the Earth and will let development keep going 64, 112

synthesis making new molecules 112

synthetic material a material that has been made artificially rather than one that occurs naturally 112

systematic something that is regular or deliberate. a systematic error is one usually due to the equipment which always throws the readings out in the same manner, for example if the zero is incorrect on a scale then all the readings will be out by the same amount 166

T

tendons at the ends of muscles; they attach muscles to bones 30

terminal speed the final speed reached by a falling object if is allowed to fall until the drag force from the air balances the pull of gravity on the object 152

theory an idea to explain evidence 116

thermal decomposition reaction a reaction where a substance splits up into new substances as a result of being heated. limestone (calcium carbonate) splits up into calcium oxide and carbon dioxide if it is heated. this is an example of a thermal decomposition reaction 78

turbine this turns when you transfer kinetic energy to it from wind, water or steam 126

turning effect the effect of a force applied to a lever or something that can turn around a pivot 162

U

unbalanced force opposing forces that are unequal 146

unethical not ethical; not matching accepted moral standards 32

unit a standard for measurement; examples are mm, g, Pa, N 142, 160

V

variation differences 2, 14

visual something that can be seen; representing something with shapes or diagrams, possibly including colour; a pie chart is a visual method of presenting data rather than putting it in a table 128

vitamins nutrients that we need in small amounts to stay healthy 26

voltage a measure of the amount of energy supplied to an electric circuit 122

voltmeter device used to measure voltage 122

W

watt unit of power; there are 1000 watts in a kilowatt (kW) 124

weed a plant growing in the wrong place 60

weedkiller a chemical that kills weeds; also called a herbicide 60

weight the force of gravity on a mass; measured in newtons 132

withdrawal symptoms the effects on the body of not getting a drug on which it is dependent 28

Y

yield how much food or other useful material a plant or animal crop can produce 58, 60, 64

Acknowledgements

AH Marks & Co Ltd 9H.3D; **Alamy** 9A.3A (Arco Images), 9C.4A Graham Morely, 9E.1C, 9F.4J (ImageState), 9F.1A (Daniel Templeton), 9F.4B (The Natural History Museum), 9F.4K (Leslie Garland Picture Library), 9G.4A (Paul Glendell), 9H.3F (Jupiterimages/Agence Images), 9H.3H (Photofusion Picture Library), 9H.4A (Jupiterimages/Goodshoot); **Andrew Lambert Photography** 9E.1H, 9E.1I, 9E.3A, 9F.1B, 9F.1C, 9F.1D, 9F.3A, 9F.3B, 9F.3C, 9F.3D, 9F.3E, 9F.4C, 9F.4D, 9F.4E, 9F.4F, 9F.4G, 9H.3A, 9H.3B, 9H.3C; **Cephas Picture Library** 9A.1B (Ted Stefanski); **The Board of Trustees of the Royal Botanic Gardens, Kew**; 9A.HSW.A; **Corbis** 9A.1D (Henry Diltz), 9A.3B (Yann Arthus-Bertrand), 9A.3D (Jim Craigmyle), 9A.3F (Ashley Cooper), 9A.6A (Najlah Feanny), 9B.1A (Lawrence Manning), 9B.1B (Ray Morsch), 9B.1C (David Pollack), 9B.1E (Tom & Dee Ann McCarthy), 9B.1F (Tim Kiusalaas), 9D.3C (Danny Lehman), 9D.5A (Nik Wheeler), 9D.5B (Julia Waterlow/Eye Ubiquitous), 9E.4A, 9E.4G (Ted Speigel), 9E.4D (Michael Freeman), 9F.2E (Galen Rowell), 9F.4H (Michael S Yamashita), 9G.1F (Pat O'Hara), 9G.3B (Document General Motors/ Reuter R/Corbis Sygma), 9I.1A (Miles/Zefa), 9K.1A (Tom Brakefield), 9K.1B (Rudi von Briel), 9K.1C (Corbis Sygma), 9K.5C (Peter Casolino); **DK Images** 9F.3F (Dean Belcher/Dorling Kindersley); **Ecoscene** 9G.2A (Erik Shaffer), 9G.2D (Nick Hawkes), 9G.3C (Vicky Coombes); **Eye Ubiquitous** 9G.1G (Sarah Errington); **FLPA** 9D.2C, 9D.3A, 9D.3B, 9D.3D, 9D.5C (N. Cattlin), 9E.4E (Inga Spence); **Fischer Scientific** 9E.HSW.A; **Fodder Solutions** 9D.5D; **Galaxy Picture Library** 9F.2A; **Getty Images** 9K.6A (AFP); **Nature Picture Library** 9A.4C (Kim Taylor); **PA Photos** 9K.5A (Neil Simpson); **Paul Beard Photo Agency** 9K.4A; **Photolibrary** 9D.1A (Sian Irvin/Fresh Food Images); **Reuters** 9I.HSW.A (Remy Steinegger); **Sam Ellis** 9I.2A; **Science Photo Library** 9A.1A (Gary Parker), 9A.1C (Alex Bartel), 9A.1E (Bluestone), 9A.3C (Rosenfeld Images), 9A.3E (John Heseltine), 9A.4A (Bjorn Svensson), 9A.4B (Peter Chadwick), 9B.1D (Erika Craddock), 9B.2A, 9B.2B (Matt Meadows, Peter Arnold), 9B.4A (St. Mary's Hospital Medical School), 9B.4B (Biophoto Associates), 9B.6A (Damian Lovegrove), 9B.6B (Gaillard Jerrican), 9B.6C (Connor Caffrey), 9D.3E (David M Campione), 9D.4A (BW Hoffman), 9D.HSW.A (Martin Bond), 9E.1A, 9I.4A (Sheila Terry), 9E.1B (Lawrence Lawry), 9E.1D (Erich Schrempp), 9E.1E, 9F.4A (George Bernard), 9E.1F, 9F.4I (Charles D Winters), 9E.1G (Simon Lewis), 9E.4C (Dr Jeremy Burgess), 9E.4H (Peter Menzel), 9F.2C (Maximilian Stock), 9F.4L (Alan Sirulnikoff), 9G.1D (Leslie J Borg), 9G.1E, 9H.3G (Martyn F Chillmaid), 9G.2B, 9H.1A (Adam Hart-Davis), 9G.3A (Wesley Bocxe), 9G.4B (Dan Guravich), 9G.HSW.A (DA Peel), 9G.HSW.B (D Phillips), 9H.1B (Magrath Photography), 9H.3E (Dr Mark J Winter), 9H.3I (Dr Morely Read), 9H.3J (Geoff Kidd), 9H.3K (Rosal), 9H.HSW.B, 9H.HSW.C (SPL), 9J.2A (Rev. Ronald Royer), 9J.3A (NRSC), 9J.3B (NASA), 9K.5B (Takeshi Takahara); **Shutterstock** 9G1.A (Victor Newman), 9K2.A (Chrisfoto); **Still Pictures** 9H.1C (David Woodfall); **The Allan Cash Picture Library** 9K.2B; **The Advertising Agency** 9B.5A; **Vanessa Miles** 9D.2A, 9D.2B, 9E.4B, 9E.4F, 9F.2B, 9F.2D, 9G.1B, 9G.1C, 9G.2C, 9H.HSW.A, 9I.3A, 9I.4B.

Image references show the Unit and Topic of the book (eg. 9A.1) and the order of the image in the Topic from top to bottom, left to right (e.g. 9A.1B is the second photograph in Topic 1 of Unit 9A).

Series advisors Andy Cooke, Jean Martin
Series authors Sam Ellis, Jean Martin

Series consultants Diane Fellowes-Freeman, Richard Needham

Based on original material by Derek Baron, Trevor Bavage, Paul Butler, Andy Cooke, Zoe Crompton, Sam Ellis, Kevin Frobisher, Jean Martin, Mick Mulligan, Chris Ram